FACILITATING POWER TRADE IN THE GREATER MEKONG SUBREGION

ESTABLISHING AND IMPLEMENTING A REGIONAL GRID CODE

DECEMBER 2022

ASIAN DEVELOPMENT BANK

Contents

Tables, Figures, and Map

Tables

Figures

Map

Foreword

Over the past decade, the Greater Mekong Subregion (GMS) Regional Power Trade Coordination Committee (RPTCC) has made significant progress in deepening cooperation among its members to scale up power trade. The RPTCC has taken steps to harmonize power sector regulations, technical performance standards, and grid codes. It has been a privilege for the Asian Development Bank (ADB) to support this work to help GMS members reap the benefits from more optimized use of diverse energy resources across the region.

This report reflects a 5-year endeavor among government officials and representatives from utilities with the support of international experts to reach an agreement on a range of complex technical and institutional issues in establishing a common Regional Grid Code (RGC) for the GMS. Power planning studies have confirmed that significant savings can be achieved, particularly through reserve sharing, load shifting, and joint use of variable renewable energy and hydropower resources. The report addresses the technical and operational issues of interconnections along with the steps that are required to harmonize national grid codes within an overarching regional structure of the RGC. The development of the proposed RGC considered experiences from global power markets and worldwide advancement in high and super-high voltage transmission technologies and reflected the specific power utility development conditions and stakeholder interests of GMS members.

In 2020, ADB published *Harmonizing Power Systems in the Greater Mekong Subregion: Regulatory and Pricing Measures to Facilitate Power Trade* in hopes of fostering and encouraging cross-border power trade. This new report on *Facilitating Power Trade in the Greater Mekong Subregion: Establishing and Implementing a Regional Grid Code* follows through the progress made over the years. Both reports present a comprehensive set of rules, standards, and techniques to help enhance cooperation. The findings presented in these studies are particularly relevant to government ministry planning departments, regulatory bodies, power utilities, and potential private sector investors that play a critical role in expanding power trade within the GMS.

We share an increasingly uncertain outlook in the energy market with the impact of high and volatile fuel prices. The GMS region is endowed with large volumes of renewable resources that can enhance the region's energy security. We hope that a future harmonized grid can enable the transition toward clean energy, and benefit our communities with reliable, affordable, and sustainable electricity services.

Ramesh Subramaniam
Director General
Southeast Asia Department

Abbreviations

ADB	Asian Development Bank
AGC	automatic generation control
ASEAN	Association of Southeast Asian Nations
BESS	battery electric storage system
CEERD	Center for Energy Environment Resource Development - Thailand
CSG	China Southern Power Grid
DSO	distribution service operator
EDL-T	Électricité du Laos Transmission Company Limited
EGAT	Energy Generating Authority of Thailand
ENTSO-e	European Network of Transmission System Operators for Electricity
EPRI	US Electric Power Research Institute
FACTS	flexible alternating current transmission systems
GIZ	Gesellschaft für Internationale Zusammenarbeit
GMS	Greater Mekong Subregion
HV	high voltage (typically all voltages over 200 kilovolts)
HVAC	high voltage alternating current
HVDC	high voltage direct current
HVDC-VSC	HVDC using voltage source control
IEA	International Energy Agency
IEEE	The Institute of Electrical and Electronic Engineers
IFC	International Finance Corporation (World Bank Group)
IGA	Inter-Governmental Agreement on Regional Power Trade in GMS
IPP	independent power producer
IRENA	International Renewable Energy Agency
JICA	Japan International Cooperation Agency
JUMPP	Japan-US Mekong Power Partnership
kV	kilovolts
LFC	load frequency control
MEA	Municipal Electricity Authority (of Bangkok)
MHI	Manitoba Hydro International
MV	medium voltage (subtransmission and distribution voltages 110 kilovolts -20 kilovolts)
MW	megawatt
NCC	National Control Center
NGC	National Grid Code
NPV	net present value
NREL	National Renewable Energy Laboratory
PEA	Provincial Electricity Authority (of Thailand)

PPA	power purchase agreement
RETA	regional technical assistance
RGC	regional grid code
RPCC	Regional Power Coordination Center
RPT	regional power trade
RPTCC	Regional Power Trade Coordination Committee
SAARC	South Asian Association for Regional Cooperation
SCADA	System Control and Data Acquisition
Sida	Swedish International Development Cooperation Agency
TSO	transmission system operator
TTC	transmission transfer capability
UCPTE	Union for the Co-ordination of Production and Transmission of Electricity
V2G	vehicle-to-grid
VFT	variable frequency transformer
VRE	variable renewable energy (typically wind and solar power)
WGPG	Working Group on Performance Standards and Grid Code
WGPO	Working Group Planning and Operations
WGRI	Working Group on Regulatory Issues

Executive Summary

The Greater Mekong Subregion (GMS) comprises six politico-economically disparate countries, i.e., Cambodia, the Lao People's Democratic Republic (Lao PDR), Myanmar, Thailand, and Viet Nam, which are members of the Association of Southeast Asian Nations (ASEAN), along with the Guangxi Zhuang Autonomous Region and Yunnan Province of the People's Republic of China (PRC). The GMS is one of a dozen groups of countries around the world grappling with the issues of how to foster regional electricity power trading in the rapidly changing energy sector. In most countries, power sector institutions have been traditionally self-sufficient and largely focused on maintaining secure and reliable national power networks in the absence of a competitive power market. In its capacity as the GMS Secretariat, the Asian Development Bank (ADB) has been supporting several sectors of the region's economic development since 1992. It is recognized by its development partners as the lead agency in coordinating technical, financial, and institutional support for developing an open access transparent power trading regime within the GMS.

This report explains ADB's supporting role in coordinating and encouraging the GMS Transmission System Operators (TSO), who manage the respective national electricity high voltage networks, to establish the Regional Power Trade Coordinating Committee (RPTCC) and implement an agreed Road Map for grid-to-grid power trading to start after 2022. Under the first stage of the Road Map, the TSOs agreed on a range of complex technical and institutional rules for power trading that are now incorporated in the GMS Regional Grid Code (RGC) that was recently published on the GMS website. Two recent energy planning studies demonstrate how the GMS regional power trade can bring short-term net economic benefits of about 5–8 times the investment worth from $3 billion to $4 billion in about 14 proposed GMS interconnections.

The aim of the next stages of the Road Map is to both harmonize the existing national grid codes (NGCs) to comply with the RGC's overarching regional structure and build a series of interconnections to enable fully synchronized power trading throughout the GMS before 2030. However, there is still much to do to implement a grid-to-grid power trading regime, including the construction of new transmission lines and substations dedicated to cross-border trading. Given the uncertainties of regional politics, along with the concerns about the impact of coronavirus disease (COVID-19) on electricity demand and supply within the GMS, it is clear that the final stage of the Road Map is some way off. This was envisaged at the time, with ADB's *Greater Mekong Subregion Energy Sector Assessment, Strategy, and Road Map* published in 2016 noting that "most of the GMS countries will have moved to multiple sellers–buyers regulatory framework, so a wholly competitive market can be implemented.

Moreover, until the necessary central power trading organization is established and permanently staffed by members from the GMS, it will be difficult for the RPTCC with its part-time members to implement the plans to fully synchronize and harmonize the GMS transmission networks within the next decade. In the meantime, working groups must be empowered by their respective GMS authorities to continue to undertake the necessary tasks of gaining regulatory approval of the RGC for its further development and implementation, performing detailed feasibility studies of interconnections, promoting the development of efficient electricity markets, facilitating information exchange between TSOs, stakeholders and regulators, and providing statements of opportunity for investors.

Institutional Planning for Regional Power Trade

With a strong commitment to regional power trading from all six member countries, the Intergovernmental Agreement (IGA) on regional power trade was signed in 2004. This established the RPTCC along with its two working groups, i.e., the Working Group on Regulatory Issues (WGRI) and the Working Group on Performance Standards and Grid Code (WGPG). Subsequently, two intergovernmental memorandums of understanding (MOUs), known as MOU-1 and MOU-2, provided the basis for the RPTCC to commence its initial work program. The third MOU, dated December 2012, authorized the establishment of the Regional Power Coordination Center (RPCC). The MOU also formally recognized the RPTCC as the coordinator of the GMS objective to promote the synchronized power system operations toward a unified, fair, and transparent regional electricity market with stable and reliable electricity supply at the most economic cost. While agreement on the location of its RPCC headquarters is yet to be reached, the RPTCC working groups will continue building the foundations for the eventual move to a regional power market.

It has been particularly challenging for the RPTCC working groups to reach an agreement on common technical and commercial standards for power trading through cross-border synchronous operations. To support the GMS countries and the RPTCC working groups, ADB provided a regional technical assistance program (RETA-6440). On completion of the project in 2012, there was an effective presentation to the regional regulators to explain the outcome of the studies that helped to maintain the support of the respective GMS government bodies. In June 2016, ADB prepared its Greater Mekong Subregion Energy Sector Assessment and Road Map (hereinafter referred to as the Road Map; Appendix 1). ADB also provided further technical assistance to the RPTCC (RETA-8330) to support the successful development of technical, operational, and market codes. Subsequently, in March 2019, the GMS RGC was accepted as a reference document and later published on the GMS website. It was also agreed to expand the tasks of WGPG and rename it as the Working Group on Planning and Operations (WGPO) to better concentrate its efforts on developing an RGC implementation program. Two studies of the benefits of power trade were also completed that confirmed the benefits of power trading and identified an optimum plan for grid-to-grid interconnections as described in the next section.

ADB's leadership, the flexibility of its technical assistance program (Appendix 2), and assistance from development partners have been essential in reaching a consensus on the development of a regional power market in the GMS. Partners include the Swedish International Development Cooperation Agency (Sida) and Agence Française de Développement, the Department of Foreign Affairs (Australia), the World Bank, and the Japan International Cooperation Agency (JICA). The involvement of other development partners, including organizations providing general technical and institutional advice, is summarized in Appendix 3.

Networks and Stakeholders

Due to the geography and uneven energy resources of the GMS, there are significant differences in the design and operational characteristics of the transmission networks in each member country. Although there are several cross-border interconnections within the GMS, these were designed to support independent power producer projects and not to promote grid-to-grid synchronous power trading to benefit the whole region. There is no agreement among GMS countries on how a 500-kilovolt (kV) or high voltage direct current (HVDC) backbone transmission structure might eventually develop to link potential resources and demand centers in an optimal way. This is partly due to the differences in planning priorities, TSO concerns regarding the risks of power system instability, and political concerns about the impact of power trading on the security and development of national resources.

The China Southern Power Grid (CSG) is by far the largest transmission network in the GMS and should be able to be interconnected from key 500-kV substations to the other five ASEAN member grids without significantly impacting its design. The two largest power sector organizations among the other members of the GMS (Thailand and Viet Nam) have extensive 500 kV networks. They also have radial connections to power stations using coal and hydropower resources located in neighboring Lao PDR. The power sectors in Cambodia, the Lao PDR, and Myanmar were less developed but have significant potential hydropower resources that can be profitably deployed within the GMS. For many years, the Lao PDR has been proactively developing policies and projects for hydro exports under long-term power purchase agreements (PPAs). Moreover, the Lao PDR is the only GMS country with contiguous borders with all the others. It is thus well-suited to having a common 500-kV backbone grid and direct fiber-optic communication links that could enable pooling hydro reserve capacity to support expansion of GMS variable renewable energy (VRE) resources. Recently, the Lao PDR and CSG formed the joint venture transmission company EDL-T that proposes a hybrid high voltage alternating current (HVAC)/high voltage direct current (HVDC) backbone grid through the Lao PDR, with facilities to connect with Cambodia, Thailand, and Viet Nam.

The RPTCC plans for the next stage of development to embark on grid-to-grid power trading needs to consider how best to include the interest of all future power market stakeholders (i.e., regulators, owners, financiers and operators of generation assets, operators of transmission and distribution systems, manufacturers, installers, and electricity consumers). It will also need to maintain the role of ADB and its development partners to provide continuing technical and institutional advice and to finance priority investments that may not be of interest to developers. In the interests of an efficiently coordinated development of the power market, the RPTCC plans for further enhancement of the grid codes to be coordinated within a permanently established RPCC facility to manage the interests of multiple individual stakeholders in accordance with the requirements of their national regulators.

Interconnection Options

A typical grid-to-grid interconnection arrangement is normally developed by analyzing load flows, prospective fault levels, and the stability of both the connection and the combined grids under a variety of circumstances. Key issues of concern include thermal limits, stability limits, frequency control, and voltage regulation, which are the main constraints on transmission line operation. Other transmission concerns include loop and parallel path flows, available transfer capacity for wheeling purposes, and undesirable reactive power flows. There can also be system-wide issues that may adversely impact stakeholders even if they are not directly involved in specific grid-to-grid transactions. These include the coordination of planning, operations and maintenance, the impact of power transfers through systems that are aging or in poor repair, and the operation of large nuclear power plants prone to delays in restarting after sudden forced outages.

HVAC and HVDC transmission interconnections have cost and technical advantages depending on their application. For short distances, HVAC interconnections are generally preferred for their lower cost and compatibility with existing grids. However, the use of HVAC requires both grids to be synchronized and care must be taken to ensure the resulting interconnections do not give rise to a variety of complex problematic power flows and instability issues. Grid-to-grid system disturbances that adversely affect the quality of power can be mitigated using technologies such as flexible alternating current transmission systems (FACTS), inverters, and variable frequency transformer (VFT) technologies. Such devices can also be used to significantly reduce the impact of VRE and the harmonic disruptions caused by large industrial loads. HVDC interconnections have traditionally been used for power transfers greater than 500 kilometers, particularly through unpopulated regions. However, they can also be used even on shorter distances to provide complementary technical and control capability to limit the impacts of problems in one grid being transmitted to the other. There are emerging applications using HVDC Voltage Sourced Converter (VSC) technologies that can be designed to provide "black starting" and torque synchronizing capability along with competitive ancillary services that can be used by TSOs to replace more expensive conventional reserves.

It is expected that a GMS regional grid will eventually be synchronously interconnected using a hybrid combination of HVAC/HVDC technologies. This is a general trend around the world and the GMS can be flexible in how it develops its interconnections while avoiding short-term measures (e.g., building HVDC back-to-back terminals and terminals that have no interconnecting transmission lines) that can become stranded assets. The use of HVDC interconnections to initiate grid-to-grid power trading for reserve sharing may well be the best way of prioritizing investments to meet the long-term goal of regional synchronization of GMS HVAC networks. However, only the PRC (with its extensive HVDC networks and leading technologies) and Thailand (with its HVDC back-to-back connection with Malaysia) have any experience with this transmission interconnection technology.

Grid Code Types and Applications

All grid codes are agreed sets of rules, specific to the safe and reliable connection and disconnection of two or more electrical entities, designed to enhance the security and economy of the production, transport and consumption of electricity. There are over 400 types of grid codes in 65 countries. Mostly, they are designed as national grid codes (NGCs) that enable a country's regulators, transmission and distribution operators, generators, suppliers, and consumers to function more effectively across the sector. NGCs broadly cover transmission and distribution interconnections, with a growing number of specific applications such as large wind and solar farms along with applications relating to the use of FACTS technologies and grid battery devices.

Typically, RGCs needed to govern the technical aspects of grid-to-grid power trading cover (i) the defined categories of generator connections that relate to their ability to contribute to system stability; (ii) interconnections with large industries that can feed undesirable harmonic disturbances back into the grid depending on the characteristics of the demand; and (iii) HVDC systems. They also need to cover several operation and market rules to ensure stakeholders can participate in an open and transparent regional power trading market and address issues of concern at distribution interconnections that were designed for one-way power flows from the transmission system to consumers. As VRE and distributed battery systems are installed in distribution networks (e.g., V2G or vehicle-to-grid systems), two-way power flows will become more common.

At this stage of RGC development, the RPTCC has focused on the technical issues of HVAC interconnection including the harmonization of the respective NGCs, the development of an ancillary services market, the establishment of common reliability standards, and the provision of system flexibility to incorporate increasing VRE into its existing mix of generation. In this regard, the European Network of Transmission System Operators (ENTSO-e) has provided the RPTCC with the best model of how a cohesive group of countries can all benefit from integration under a common RGC. Accordingly, it has been used as the basis for the development of the GMS RGC. The ENTSO-e codes are continuously being adapted to reflect technological changes and security concerns.

All the GMS countries have NGCs that govern how their TSOs plan and manage their power systems. Although the NGCs are continuing to be developed independently by each GMS country, they need to be compatible with an overarching and common RGC. In general terms, the RGC covers the following technical and operational issues:

(i)   **Technical considerations**. The RGC overarches NGCs, which can be more detailed to suit local regulatory requirements.

(ii)   **Provision of a glossary of terms**.

(iii) **Harmonization of NGCs**. This addresses the need for compatibility of differences between measurements, methods, procedures, schedules, specifications, or systems.

(iv)   **Reliability standards**. Planning codes are needed to ensure that problems in one region are not transferred to another region through interconnections.

(v)   **Frequency regulation**. Primary, secondary, and tertiary reserves for containment of frequency fluctuations maintain synchronous power balancing.

(vi)   **VRE sources**. Mitigate impacts of intermittency to enable load shifting and optimize the use of GMS hydro storage capability.

(vii) **System flexibility**. Modify regional transmission and distribution grid codes to enable new distributed technologies to compete in power markets.

(viii) **TSO operations**. This enables coordination between independent national TSOs and distribution system operators for load balancing and during emergencies.

(ix)   **Communications, control, and data management**. System control and data acquisition for load dispatch and emergency operations, cybersecurity, and confidentiality of data.

(x)   **Regulation and pricing**. Barriers, open access, wheeling charges, short-term trading rules, balancing mechanism.

(xi)   **Other strategic planning guidelines**.

There are emerging issues in the power market that will need to be addressed through further developments of the RGC. These relate to cybersecurity, data confidentiality, the development of grid storage, the impacts on the grid of an increasing proportion of VRE generation sources, and demand-side management technologies.

Scope of the Regional Grid Code

The GMS RGC comprises 12 subcodes designed to address technical, institutional, and market issues reflecting the characteristics of the region and accommodating the current requirements of the TSOs. The subcodes are based on common technical policies agreed by RPTCC members at the outcome of the development of the

RGC generally as established in the table of performance standards. As required by protocol, each subcode incorporates comments on specifications by the RPTCC country members that were discussed during its deliberations. In some cases, the issues raised will need to be revisited to make sure there is a common agreement with the RGC. A gap assessment of the conflicts between the RGC and the GMS NGCs is being used, along with information gathered from Cambodia and the Lao PDR to establish an RGC implementation program in the next few years.

There is a summary of internationally accepted terms, acronyms, and units used in the transmission regulations and the GMS RGC. The preamble provides an explanation of the context of the documents in terms of regional power trading, policy objectives, the relationship between the RGC and the NGCs, and the provision for a separate synchronous zone in the transition period. It provides an overview of the main sections that include the codes for governance, connection, operations, markets, and metering. There are two other complementary codes relating to TSO training and requirements for strategic planning within the region. It can be expected that additional subcodes will be needed to deal with emerging issues that are already being considered by other regional jurisdictions.

The Governance Code describes the processes to be followed in updating the RGC to improve safety, reliability, and operational standards. It sets out how stakeholders can influence the amendment process and defines who has the authority to recommend and ultimately approve and enforce changes. It elaborates on the oversight and compliance provisions that need to be observed and sets out dispute management procedures, explains how outcomes should be determined by the RPCC board, and how violations and sanctions should be administered by the RPTCC.

The Connection Code is the largest and by far the most technically specific section of the GMS RGC. It separately sets out (i) requirements for generators, (ii) HVDC connections, and (iii) demand connections. The objective is to ensure that, by specifying minimum criteria, the basic rules for connection are the same for all facilities with an equivalent capacity to influence grid operations. Countries with tighter parameters stipulated in their NGCs that fall within the bands proposed in the RGC are considered compliant. The Connection Code will enable the maintenance, preservation, and restoration of system security to facilitate optimal functioning and achieve cost efficiencies in the internal electricity market within and between synchronous areas. Each of the three specified facilities includes separate sections defining technical requirements of conventional power plants in terms of (i) frequency tolerance, active power, and frequency control requirements; (ii) voltage tolerance, voltage control, and reactive power provision; (iii) fault ride-through capability; (iv) protection requirements; (v) system restoration, islanding, and black start capability; (vi) information requirements; and (vii) connection and testing. These codes will likely need to be expanded to deal with the emerging characteristics of VRE generation characterized by low inertia and intermittent operation.

Operations Codes contain details of high-level TSO operational procedures such as demand control, operational planning, and data provision. Their main purpose is to document TSO practices so they are more transparent to stakeholders wishing to participate in power trading. These include four subcodes that focus on rules that the TSO will use for managing operational security, planning and scheduling, load frequency and reserves control, and energy and restorations. They deal with the criteria and procedures required to facilitate efficient, safe, reliable, and coordinated system operation of the GMS incorporated in the following subcodes. Each subcode is in effect a stand-alone document that duplicates many general provisions relating to regulatory aspects and approvals, recovery of costs, confidentiality obligations and agreements with TSOs not bound by the RGC. The subcodes provide a list of related terms used to monitor and manage power flows within the interconnected systems and an explanation of the regulatory and confidentially aspects of reporting relating to approvals, cost recovery, consultation, and coordination.

Market Codes contain a set of operational requirements in three subcodes for capacity allocation and congestion management, forward capacity allocation, and electricity balancing. These are codified guidelines that are an integral part of the market code family. The subcodes set out non-discriminatory rules for access conditions to the network for cross-border exchanges in electricity and rules on capacity allocation and congestion management for interconnections and transmission systems affecting cross-border electricity flows.

The Metering Code defines the metering types and functions for use at each point of exchange between grid control areas. It specifies the minimum technical, design and operational criteria to be complied with for the metering of each point of interchange of energy between control areas, TSOs, and other trading parties. The code is not concerned with the metering of connection points that are subject to NGC regulations and PPAs.

The System Operator Training Code sets out the responsibilities and the minimum acceptable requirements for the development and implementation of system operator training and authorization programs. The code defines a standard framework for operational training to provide reasonable assurance that the dispatchers have and maintain the knowledge and skills to always operate the power system safely and reliably under all conditions. The code defines the framework for operational training to provide reasonable assurance that the dispatchers have and maintain the knowledge and skills to always operate the power system in a safe and reliable manner under all conditions.

The Strategic Planning Document specifies the minimum technical and design criteria, principles and procedures for medium- and long-term development of the GMS synchronously interconnected transmission systems along with the planning data required to be shared among members. It specifies the requirements for the interchange of information between the RPCC and individual TSOs. This information enables the RPCC and TSOs to take due account of regional developments, new connection sites, or the modification of existing connection sites within the TSO's transmission network along with new or modified connections with external systems.

Currently, the GMS RGC cannot be considered a binding legal document until it is incorporated within the regulatory structure of the member countries. This will be a necessary step to encourage participation from all the governments, utilities and private sector stakeholders involved in the power market to deal with key issues of sovereignty, the economy and security of synchronous interconnections, the rapid growth and integration of decentralized renewables within their generation mix, and the deployment of new technologies to optimize the operations of their increasingly extensive and complex transmission networks.

Implementation of Power Trading

To implement the GMS RGC, it is important for the RPTCC to take responsibility for activities beyond 2022 even though there is still no agreement on the location of an RPCC facility. An interim institutional structure needs to be formulated and provided with a sufficient budget and permanent staff, if necessary, through another GMS ministerial intervention. It is vital that the respective national regulatory authorities are given the opportunity to independently review the scope of the RGC and to consider how to make it legally enforceable. Assuming the impact of the COVID-19 pandemic will prevail in the foreseeable months ahead, there is a strong case for WGRI and WGPO activities to be accelerated through virtual meetings in anticipation of the time when economic activity is restored.

In the coming months, the RPTCC needs to enforce its control over RGC implementation by designating permanent representatives from each TSO and delegating responsibilities for tasks to be completed. For example, the RPTCC homepage could be used more effectively to initiate planning and report on progress with the development of the Grid Codes. A key objective should be to involve more stakeholders by enhancing the existing RPTCC website to provide a gateway to obtaining information on GMS power trading activities. This would include links to stakeholder websites, posted opportunities for investors, details of the RPTCC working group discussions and resolutions, and so forth.

The numerous tasks agreed by the WGRI are designed to advance the development of the RGC operations and market codes by amending existing NGCs to increase the transparency of operations codes and better define short-term power trading and balancing arrangements. The implementation of a proposed pilot grid-to-grid power trading project will also help uncover many issues that still need to be addressed before power trading can begin in earnest. The WGRI must develop a strategy to renegotiate the existing PPAs with the independent power producers in consultation with national regulators. In the meantime, the WGRI should investigate cloud storage alternatives for centralizing GMS technical data that also provide a secure backup to existing national systems.

Similarly, the WGPO needs to be empowered to complete the process of aligning the technical aspects of the NGCs with the RGC while seeking stakeholder support for its enforcement before submission to the respective regulators. The WGPO needs to commence the design of the regional communications and metering systems for monitoring and managing a regional power market. It must develop a strategy for synchronization that considers alternatives such as promoting the synchronous interconnection of the ASEAN national grids before considering how the subregion should be interconnected with the much larger grid of the southern PRC. This may involve (i) investigating alternative ways of achieving interconnections including the use of HVDC-VSC and other new technologies as a means of strengthening grid-to-grid interconnections prior to using HVAC links and (ii) developing a reserve sharing project between Viet Nam and Thailand.

Conclusions and Lessons Learned

In summary, this report describes the activities of the GMS TSOs in reaching agreement on a range of complex technical and institutional issues prescribed in the RGC. It also describes the history, nature, and application of grid codes around the world, noting that there are several similar activities in process in a dozen other regional groups of international power utilities. Power planning studies have confirmed that significant savings can be achieved, particularly through reserve sharing, load shifting and the joint exploitation of VRE and hydropower resources. The methodology for allocating the network costs associated with power trading among national stakeholders has been outlined in a complementary ADB publication dealing with regulatory and pricing measures applicable to the GMS. This report addresses the technical and operational issues of interconnections along with the steps required to harmonize the use of existing NGCs within an overarching regional structure of the RGC. It considers the special characteristics of the region and the interests of its stakeholders along with global trends in technological development expected to drive the scope of future amendments to the GMS RGC.

The work to establish the RGC has been challenging for the TSO members assigned to the RPTCC working groups, often with different participants having to take time off their regular duties to undertake the tasks required. They have met more than 20 times over the last 10 years to review a variety of related technical reports, investment proposals, and gap analysis reports that will require amendments to their own NGCs. In doing so, the TSO representatives have successfully negotiated a historic international agreement on a standardized set of codes for further development. In the process, the RPTCC working groups have built up confidence in working together, identified gaps that need to be addressed and gained the support of the respective government ministries and regulators in pursuing the goal of promoting open access power trading. Both working groups need to be empowered to advise policymakers, develop, and approve wheeling charges, to enable them to continue to adapt the RGC and advise on contractual structures and technical standards. Their main aim should be to increase stakeholder participation in providing an information facility to promote transparent open access that explains the rationale for the RGC and invites submissions for further amendments to reflect the special characteristics of the GMS.

There is still a long way to go to establish a transparent open access regime of power trading in the GMS. However, without an RPCC permanently staffed by members assigned from the TSOs, it will be difficult to achieve a fully synchronized GMS transmission network before 2030. There are lessons to be learned to help the RPTCC working groups continue to deal with outstanding issues specific to the GMS including the following:

(i)   Building the institutional trust and support needed to maintain momentum with clear objectives for power trading by a target date.
(ii)   Recognizing the distractions to working groups with members who are also supporting the parallel development of their own national rapidly growing power systems.
(iii)   Accepting the important role of ADB in initiating the support of international experts to provide advice from other jurisdictions.
(iv)   Ensuring the respective regional governments and regulatory authorities are well-informed of the progress in developing the RGC and seeking their advice as to how to make it legally enforceable.
(v)   Ensuring the RGC is used to coordinate the operational differences between the management of each country's power market.
(vi)   Dealing with long-standing PPAs, sharing system data, and respecting stakeholder confidentiality.
(vii)   Effecting the first stage of grid-to-grid synchronism with a pilot project to coordinate planning and power trading with the national TSOs.

1. GMS Institutional Planning for Regional Power Trade

GMS Power Trading Opportunities

The Greater Mekong Subregion (GMS) is one of many regions in the world grappling with how to foster regional power trading in a rapidly growing sector with its institutions traditionally self-sufficient and largely focused on maintaining secure and reliable national power networks. Since 1992, power trading within the GMS has been the domain of independent power producers (IPPs). Mostly, they have developed and exported power from a generation resource in one country to a neighboring grid or customer in another under long-term bilateral power purchase agreements (PPAs).

However, other forms of cross-border power trading could be profitably introduced by GMS Transmission System Operators (TSOs) who manage national electricity high voltage networks. These would be designed to facilitate generation reserve sharing (for backup in case of grid failures), load shifting (to relieve overloaded transmission networks), and power wheeling to industrial consumers or distribution companies within neighboring grids. Such grid-to-grid power trading ventures have been popularized worldwide by the rapidly falling costs of low emission variable renewable energy (VRE) resources that will increasingly need fast-acting backup storage capability to participate effectively. Moreover, as the technical and economic viability of VRE continues to improve, power utility business-as-usual practices are facing rising public concerns about the adverse health and environmental impacts of fossil fuel emissions from large, centralized power plants.

The 2012 Road Map was developed based on a four-stage plan to achieve a fully functioning regional power market. Stages 1 and 2 of the Road Map focused on the development of a Regional Grid Code (RGC) to enable GMS grids to be synchronized through new interconnections so that development of grid-to-grid power trading could begin in 2022. Subsequently, the GMS TSOs agreed on a range of complex technical and institutional rules for power trading now incorporated in the RGC.

In its capacity as the GMS Secretariat, the Asian Development Bank (ADB) has been supporting several sectors of the region's economic development since 1992. This report explains ADB's role in coordinating and encouraging the TSOs to establish the Regional Power Trade Coordinating Committee (RPTCC) and implement a Road Map to enable grid-to-grid power trading starting in 2022. ADB has provided considerable technical assistance to this effort (Appendix 2) and is also recognized by its development partners (Appendix 3) as the lead agency in coordinating technical and institutional support for developing an open access transparent power trading regime within the GMS.

Economic Benefits of Power Trading

Over the years, there have been several economic studies of GMS energy resources showing they can be optimized by exploiting power trade opportunities within the region. The two most recent studies were financed by the World Bank (2019) and ADB (2021) with the objectives listed in Appendix 4. Both studies modeled the capital and operating costs of generation and transmission expansion plans for the GMS countries along with associated carbon emissions up to 2035 to determine the net present value (NPV) of economic benefits attributable to grid-to-grid interconnections. The World Bank study using economic assumptions prevailing in 2016, examined 18 proposed interconnections and evaluated business cases for 10 of the most promising transmission projects. The study also considered the practical issues of achieving regional synchronization to reap the benefits of fully integrated grid operations. The complementary ADB-funded study on the Manitoba Hydro International (MHI) Master Plan, reflected the coronavirus disease (COVID-19) situation better, as it examined the economic sensitivities of interconnection strategies to facilitate synchronous power trading and recommended a robust least-cost plan for their development up to 2030.

The extensive modeling used in the studies confirmed that grid-to-grid power trading will benefit all countries within the GMS. It was demonstrated that an investment of about $3 billion–$4 billion in transmission interconnections can increase the net benefits to the region by 4–6 times. These can be derived by avoiding planned thermal generation investments, deferring the need for national transmission upgrades, and lowering power generation costs. Other identified benefits include being able to deploy higher levels of renewable energy in the region, better use of hydro resources (because of diversification in hydrological conditions), and diversification in demand profiles of the interconnected countries. Both studies observed that the Lao People's Democratic Republic (Lao PDR) could play a larger role in terms of providing additional power supplies to Myanmar and Viet Nam with immediate short-term cost reductions, and over the longer term for Thailand.

Both studies concluded that the Regional Power Coordination Center (RPCC) urgently needs to develop a plan to facilitate the progression of the GMS networks toward a more tightly integrated synchronized power system. In the meantime, it is technically feasible to commence power trading using HVDC asynchronous interconnections initially to facilitate sharing reserve capacity. To mitigate the concerns of the respective country TSOs in implementing the synchronization program, it will be necessary to obtain further technical assistance (TA) to carry out extensive power system load flow, fault, and stability studies using actual system data to identify the critical power system components that would need to be brought into compliance with the RGC.

Perspectives for Implementing the GMS Power Trade

Since 2002, the respective GMS governments have signaled their unwavering support for the development of regional power trading as documented in three memorandums of understanding (MOUs). With support from ADB and other development partners, the GMS governments established the RPTCC and working groups to work on an implementation program to facilitate power trading. Eventhough an RGC and an action plan for its implementation were completed and accepted by the TSOs in 2019, the COVID-19 pandemic since 2020 hindered progress toward its implementation.[1]

The role of ADB and its development partners will continue to be critical to implementing Phases 2 and 3. The RPTCC will require substantial technical support to finalize detailed plans for implementing the RGC, including designing priority interconnection projects for financing. Some of the interconnection projects will need to be backed by sovereign guarantees since they are not likely to be of much interest to developers. Key projects should be those that can facilitate the early synchronization of GMS networks and will probably require a combination of HVDC and HVAC interconnections.

Additional support from international financial institutions could include the financing of TA feasibility studies for investments in transmission facilities, and communication links between existing control centers along with database facilities for use by the RPTCC. Other projects that will facilitate growth of the VRE sector should target the enhancement of storage capacity, including the use of battery storage system facilities near proposed wind or solar parks.

GMS Strategy for Power Trading

Intergovernmental Agreements

In 2002, the governments of the GMS signed an Intergovernmental Agreement (IGA) on regional power trade to promote development of electric power sectors in the region using appropriate technologies for mutually beneficial shared power trade and protecting the environment. The IGA specifies overall principles for cooperation requiring each member to recognize (i) regional power trade as an integral part of its sector development, (ii) the importance of achieving harmonization of technical performance parameters and standards of transmission systems and with eventual regional power trade, (iii) the desirability for open communication and information sharing, and (iv) protection of the environment by adopting appropriate technologies and plans while embarking on GMS regional power trading.

---

[1]   RPTCC. 2019. GMS Power Trade Framework. Gap Assessment and Action Plan for the Implementation of the GMS Grid Code. Ed. M. Caubet. November.

The GMS governments also set up the RPTCC, consisting of director-general level representatives from Ministries of Energy (or equivalent) from the six GMS countries. The RPTCC was given responsibility to set up short-, medium-, and long-term initiatives aiming at fulfilling regional power trade within a specified time frame and to identify and coordinate priority measures to implement regional power trade including financing means.

Because of their different levels of power system infrastructure, regulatory functions, and technical capacity, the GMS Member States agreed to a phased approach to enhancing regional power trading that would develop gradually into a multi-seller or multi-buyer open market. Toward that end, two MOUs were signed by GMS governments. MOU-1 was signed in 2005 in Kunming, PRC, and MOU-2 in 2008 in Vientiane, Lao PDR. These MOUs set the objectives and time frames for advancing GMS regional power trade.

Establishment of RPTCC

In fulfilling its responsibilities, the RPTCC prescribed two initial stages of GMS power cooperation. Stage 1 corresponds to the situation before a regional power transmission network is set up where country-to-country transactions are used to enable grid-to-grid power trade between any pair of member countries. During this period, cross-border transmission lines would be mostly associated with PPAs between countries, or with an IPP located in one country selling electricity to a national power utility in a neighboring country. The cross-border power trade during Stage 1 refers to opportunity trading using the excess capacity of the existing transmission networks among countries. Stage 2 corresponds to the time when trading will be possible between any pair of GMS countries, eventually using transmission facilities of a third regional country, noting that in this stage the available cross-border transmission capacity is limited and based on the surplus capacity of lines linked to PPAs.

The GMS countries recognized that building a mutually beneficial regional power market will involve building both physical infrastructure and an institutional framework to create conditions for an expanded interconnection between GMS power systems before moving toward an integrated power system. They recognized that regional power trade will develop gradually through institutional changes and infrastructure development that in effect transform isolated bilateral cross-border transmission projects into a regional power market through regional power systems planning; establishing institutional and regulatory arrangements; creating trading rules, codes, and standards; and building human capacity to administer day-to-day operations of future power systems. Cooperation in power trading since 2005 corresponds to conditions where cross-border high voltage power transmission was already connected between member countries (e.g., from the PRC to Viet Nam, Viet Nam to Cambodia, the Lao PDR to Thailand, the Lao PDR to Viet Nam, the Lao PDR to Cambodia, Myanmar to the PRC, etc.). In areas near national borders, the GMS countries also import and export electricity via lower voltage levels using barter trade.

MOU-2 prescribed a road map to be carried out by the RPTCC to fully achieve the realization of GMS regional power trade at Stage 1 and moving toward cross-border power trade corresponding to Stage 2. MOU-2 required the RPTCC to set up working groups to carry out a series of activities in parallel to (i) complete the study on a GMS performance standard for new regional interconnections, and (ii) establish synchronized operation for interconnected grids and the transitional arrangements to achieve GMS performance standards. The sequence of activities includes plans to complete the following activities:

(i)   Conduct a study on transmission regulations to coordinate the operation and power flow control in grid-to-grid interconnection synchronization and operation.
(ii)   Prepare an indicative power interconnection master plan and propose new priority interconnection projects for undertaking feasibility studies.
(iii)   Conduct a study on regional metering arrangements and communication systems in grid-to-grid interconnection for implementation during Stage 1.
(iv)   Conduct a study on power trade rules, including dispute resolution outside the existing PPAs for implementation during Stage 1.

In preparation for moving to Stage 2, the working groups would have to

(i)   Study the necessary GMS grid codes (operational procedures), which include performance standards, metering, communication, and coordination procedures for regional network interconnections.
(ii)   Conduct a study on third-party access to interconnections, giving priority to contracted PPAs.
(iii)   Conduct a study to identify regulatory barriers to the development of power trade and implementation.

Between 2002 and 2012, with the signing of the Intergovernmental Agreement and the establishment of RPTCC power cooperation and trade in the GMS, the region saw significant infrastructure development. As member countries recognize cross-border power trade as an integral part of each member's electric power sector, dialogues have been conducted, planning has been coordinated, and investments have been called in, resulting in a leap in the number of cross-border transmission interconnections and their transmitting capacity.

The number of member countries taking part in cross-border power trade has similarly increased. Prior to 2002, countries traded electricity only in the areas near national borders at medium voltage levels (e.g., 22 kilovolts [kV] and below). Only the Lao PDR and Thailand engaged in power trade at the 230 kV level via two transmission links with a total capacity of 560 megawatts (MW). By 2012, 14 cross-border transmission lines were in operation among all six GMS countries with a total traded capacity of 4,030 MW. Power trade capacity reached 730 MW between the PRC and Viet Nam, 2,560 MW between the Lao PDR and Thailand, 540 MW between Myanmar and the PRC, and 200 MW between Cambodia and Viet Nam. The volume of GMS two-way regional power trade in 2010 stood at 34,100 gigawatt-hour (GWh). Since 2012, cross-border power trade has been characterized by slower infrastructure development but much stronger institutional development. By 2020, five more cross-border transmission interconnections had been built, raising the total trade capacity in the GMS region to 8,870 MW. Notably, this period saw the addition of 500 kV transmission lines for large volumes of power trade. Two-way GMS regional power trade increased to 37,500 GWh in 2014 and roughly 91,000 GWh in 2019.

Working Group Activities

In accordance with the requirements of the third MOU (2012), the RPTCC determined the composition of and the chair for the Working Group on Performance Standards and Grid Code (WGPG).[2] This comprises one designated representative per country: (i) from the national TSOs, or national power authorities or utilities performing the functions of a TSO with primary responsibility for transmission planning and operation, (ii) from power authorities or national power utilities performing the functions of a power producer with primary responsibility for generation planning and operation, and (iii) as a ministry official with primary responsibility for regulatory aspects of the electricity sector or a representative from the national regulatory agency. The designated representatives were given sufficient capability and full mandates from their respective institutions to make all relevant decisions.

To facilitate and promote the synchronized operation of national power system operations toward a unified, fair, and transparent regional electricity market, the following issues need to be addressed collectively by the GMS Member States to establish a harmonized regulatory framework: pricing, a legal framework for third-party access to the grid and wheeling obligations, capital account and current account convertibility, a legal and regulatory basis for commercial transactions addressing currency exchange issues, and procedures for resolving disputes fairly and efficiently. Toward this purpose, some urgent actions have already been identified, such as removing the existing regulatory barriers for cross-border trading and creating real market openness.

The RPTCC established the Working Group on Regulatory Issues (WGRI) and the WGPG to support the development of the GMS regional power market at the 11th meeting of the RPTCC held on 9 and 10 November 2011 in Ho Chi Minh City, Viet Nam. The GMS member countries have defined the responsibilities of each working group. The scope of activities expected to be performed by the WGPG include the following:

(i)   Conduct review of and provide recommendations on common performance standards for satisfactory operational security, reliability, and quality of supply, for approval and adoption by the RPTCC.

(ii) Establish and offer recommendations on a regional grid code for approval and adoption by the RPTCC. The regional grid code shall cover planning,

---

2   The MOU is dated 12 December 2012 and signed by the respective power and energy ministers of Cambodia, the Lao PDR, Myanmar, and Thailand along with the Vice Minister of Viet Nam and Deputy Administrator of the National Energy Administration of the PRC.

connection, operation, interchange scheduling and balancing, data exchange, metering, and system operator training.

(iii) Monitor the implementation and enforcement of performance standards and the regional grid code.

(iv) Report the work progress of the program to the RPTCC.

(v) Submit position papers and recommendations on all matters related to power system reliability and safety for discussion and approval by the RPTCC.

(vi) Define any complementary TA and training programs needed.

(vii) Perform other functions and activities as assigned by the RPTCC.

Likewise, the RPTCC established the WGRI comprising one designated member per country from (i) each ministry with primary responsibility for regulatory aspects of the electricity sector or a representative from the national regulatory agency, (ii) the respective national TSO or national power authorities or utilities performing the functions of a TSO with primary responsibility for regulatory and market issues, and (iii) national power authorities or national power utilities performing the functions of a power producer with primary responsibility for regulatory and market issues. The designated representatives shall have sufficient capability in addressing institutional, legal, and commercial issues and a full mandate from their respective institutions to make all relevant decisions. The scope of activities to be performed by the WGRI include the following:

(i) Review the regulatory and commercial barriers named in the GMS members and recommendations on how to remove these barriers, for discussion and approval by the RPTCC.

(ii) Conduct review of and provide recommendations on the GMS market design and the steps toward the creation of a unified, fair, and transparent GMS regional electricity market for discussion and approval by the RPTCC.

(iii) Submit position papers and recommendations on issues related to the structure of the electricity sector at the national level for harmonization and adoption of a common target in terms of market opening, such as restructuring (unbundling of TSOs from generation and supply), establishment of market rules, access tariff method, market monitoring, etc., for discussion and approval by the RPTCC.

(iv) Offer recommendations for harmonization of the national regulatory functions at the regional level for approval by the RPTCC.

(v) Report the work progress to the RPTCC.

(vi) Define complementary TA and training programs needed.

(vii) Perform other functions and activities as assigned by the RPTCC.

Role of ADB

ADB has been supporting GMS regional power cooperation since its start in 2002. ADB has served as the Secretariat for GMS regional power cooperation and has provided important TA support to the RPTCC and its working groups to achieve IGA goals. Appendix 2 lists the scope of the ADB TA programs for power trading, along with a list of the relevant knowledge products that have contributed to the implementation of the GMS strategy.

In 2007, after the GMS countries signed MOU-1 in 2005, ADB prepared a regional TA for Facilitating Regional Power Trading and Environmentally Sustainable Development of Electricity Infrastructure in the Greater Mekong Subregion. This was implemented between 2007 and 2012 to address the environmental aspects of regional power interconnections and trading in the GMS. The TA supported an intergovernmental MOU for establishing an RPCC dedicated to GMS power trade intended to be a major step toward realizing a regional power market. In 2010, this TA updated a regional power master plan (first prepared in 2006) and charted a mechanism for sharing the benefits of regional power trade in a fair and transparent manner.[3]

In the past, ADB had financed several in-country subtransmission investments in GMS countries, for example, in the Lao PDR along with 500 kV in transmission investments to support the development of the Lao PDR hydro export projects to Thailand and Viet Nam.[4] ADB is currently considering proposals for three more cross-border transmission investments between the Lao PDR, Thailand, and Viet Nam. The Lao PDR–Viet Nam project is, however, expected to be designed for bilateral power exports but needs to be considered by the RPTCC in terms of how they might form part of a future grid-to-grid power trading regime. In this respect, regional power trade through HVAC synchronous connections necessitates integrated national power systems that will need a high degree of technical compatibility and careful system planning and operational coordination to minimize the threat of voltage collapse, dynamic and transient instability, or supply disruptions.[5]

In support of the two working group work programs, in December 2014, ADB supplied a second regional TA program for Harmonizing the Greater Mekong Subregion Power Systems to Facilitate Regional Power Trade. The grant funding was used to provide expert advice to the working groups to develop the RGC and its associated tariff method. It was used to recruit several technical experts with experience in system planning, grid code preparation, power generation planning, and regulatory experience. At the end of his assignment, the lead ADB consultant prepared a gap analysis to identify areas where there were differences between the national grid codes (NGCs) and the RGC. This information was needed to help plan the RGC implementation and to continue with its development to deal with changes in technology and market requirements.
The TA also funded the GMS Regional Transmission Master Plan completed in 2021.[6] The MHI Master Plan developed a set of generation and transmission development

---

[3]   After the TA was completed, a comprehensive presentation of the outputs was made to the officers of the Energy Regulating Commissions. T. Lefevre. 2012. Facilitating Regional Power Trading and Environmentally Sustainable Development of Electricity Infrastructure in the GMS. Presentation to Energy Regulating Commission. 12 October.

[4]   ADB. 2020. GMS Regional Investment Framework 2022 prepared for the 27th RPTCC meeting. 15 October. Source: ADB. 2020. Summary of Recommended Actions Described in the RGC Consultant's Progress Report V8: Section 4 WGPO Tasks 5-10.

[5]   These issues were also identified at the RPTCC 27 meeting held on 15 October 2020.

scenarios to account for the uncertainty related to demand growth, and economic and technological factors. Its final recommendations considered current fuel prices, and anticipated cost reductions in VRE generation along with the potential for other technical developments such as battery storage, nuclear power, and cross-border power trade. Based on the existing and potential interconnections, the recommended transmission interconnections are much more extensive, indicating that by 2035 there will be integrated 500 kV and 230 kV synchronous connections with some HVDCs where it is considered advantageous to do so.

Working Group Achievements

Building on the achievements and recognizing the lessons of past initiatives, the TA projects were intended to intensify support to GMS power systems while working more closely with GMS countries and development partners. They specified the following outputs: (i) continued support to RPTCC activities, including the establishment and start of operations of the RPCC; (ii) continued support to WGPG activities, including consolidating identified gaps and proposed remedies from each GMS member country and making recommendations regarding an implementation plan to meet regional performance standards and grid codes; (iii) completing the study on transmission regulations and making recommendations regarding a three-tiered protocol for technical coordination; (iv) completing the study and making recommendations on standard regional meeting arrangements; (v) continued support to WGRI activities, such as providing strategic guidance to WGRI in setting up a system of power trading in the GMS; (vi) completing and making recommendations regarding the guidelines for regulatory framework improvement for mid-term GMS cross-border power trading; and (vii) completing the study on and making recommendations regarding GMS transmission market and pricing mechanisms. Members of RPTCC and two working groups collectively completed most tasks specified under MOU-2 (Table 1).

In March 2019 at RPTCC 25, the draft GMS RGC was accepted by GMS members as the official reference document recognizing that further changes would be an ongoing requirement to resolve any differences that might better reflect the characteristics of the GMS region. It was published on the ADB GMS website in January 2021.

After successfully completing Stage 1 objectives stated in the GMS Road Map, Stage 2 was expected to begin. The *Greater Mekong Subregion Energy Sector Assessment, Strategy, and Road Map* published by ADB in 2016 notes that Stage 2 would begin with the development of a strategy "to implement grid-to-grid power trading between any pair of GMS countries using the transmission facilities of a third regional country." The newly renamed Working Group on Planning and Operations (WGPO) was asked to focus on activities to enable the implementation of the RGC. This would also require the WGPO to consider how HVAC interconnections, particularly those between Cambodia and the Lao PDR, could be designed to form part of a wider 500 kV synchronous grid.

---

[6]   ADB. 2020. *Regional Power Master Plan: Harmonizing the Greater Mekong Sub-Region (GMS) Power System to Facilitate Regional Power Trade.* Consultant's report. Manila.

Table 1: Summary of Working Group Outputs to Enable GMS Power Trading

WGPG	Objective	Outputs (as described in this report)
Performance standards	GMS performance standards proposed, September 2016	GMS standard for frequency GMS standard for voltage GMS standard for harmonics GMS standard for maximum fault clearing time GMS standard for planning studies
Transmission regulation	Study on transmission regulation completed. Recommendation for adoption, August 2017[a]	1. Policy on Scheduling and Accounting 2. Policy on Coordinated Operational Planning 3. Policy on Communication Infrastructure 4. Policy on Data Exchange, Rules for Handling of Data, Code of Confidentiality 5. Policy on Load Frequency Control and Reserve
Develop RGC	GMS RGC	Proposed for adoption, March 2019; published January 2021, Implementation Strategy, November 2019.
Update MHI Master Plan	Interconnection Strategy	GMS transmission master plan updated, February 2021
WGPG	Objective	Outputs (incorporated in the ADB knowledge product on tariffs and wheeling)
Harmonization	Reforms to harmonize the electricity sector across GMS member countries.	Study on regulatory barriers to the GMS regional power trade completed, March 2016. Major regulatory barriers per country identified. Recommendations for overcoming regulatory barriers proposed in March 2016.
Fair and transparent transmission tariff methodology	Study on third-party access and methodology for transmission charge using third-party transmission assets.	Third-party access provisions April 2017. Wheeling charge methodology. Review on international practices on transmission costing and tariff methodologies. Tariff methodology for GMS proposed. Application of GMS methodology for calculating wheeling charge of practical cases.
Bilateral short-term trading	Study of available transmission capacity.	Evaluation of available transfer capacity December 2017. Bilateral trading rules. Reconciliation of bilateral trade imbalances. Financial settling rules.

ADB = Asian Development Bank, GMS = Greater Mekong Subregion, MHI = Manitoba Hydro International, RGC = regional grid code, RPTCC = Regional Power Trade Coordinating Committee, WGPG = Working Group on Performance Standards and Grid Code.

[a] RPTCC member comments on the Transmission Regulations Policies 1 to 4 are summarized in Attachment 4 RPTCC 24. Comments on Grid Codes are incorporated with the respective Codes.

Source: ADB. 2018. *WGPG on Performance Standards and Regional Grid Code* prepared for the 24th RPTCC meeting. 18–20 June.

Regional power trade necessitates regionally integrated power systems, which require a high degree of technical compatibility and careful system planning and operational coordination to minimize the risk of voltage collapse, dynamic and transient instability, or supply disruption. In the absence of such coordination and compatibility, multiple systems across member countries may be downed by cascading outages arising from technical or other faults that originate in a single country.

The presence of varying organizational frameworks, technical capabilities, and even cultural distinctions can all contribute significantly to supply interruption. To make further progress on regional power trade and accelerate progress from Stage 1 to Stage 2, GMS members must do much more to realize the full benefits of synchronous operations. In this regard, the establishment of the RPCC will demonstrate ownership and leadership by members of the regional power trade and market development process. The institution must have a legal identity and be fully dedicated to managing cross-border power infrastructure and trade in the GMS. The continuing involvement of the WGPG and WGRI will help bridge gaps between GMS country technical standards and the regulatory framework to enable a regional trade in power by harmonizing (i) performance standards and grid codes that establish technical rules for the coordinated planning and operation of the regional electricity market; and (ii) regulatory frameworks, pricing, legal frameworks for third-party grid access, and wheeling obligations. These prerequisites are fundamental to the construction of a unified, fair, and transparent regional electricity market.

Implementation of Regional Grid Code

The following subsections of this report outline the considerations for the further development and implementation of the RGC relating to (i) the characteristics and planned developments in the GMS power systems, including information on recent institutional developments in the Lao PDR, and the identification and interests of stakeholders in the development of the grid codes for power trading; (ii) the technical alternatives for interconnections between synchronous and asynchronous systems; (iii) the applications that the GMS RGC and the respective NGCs are designed to address; and (iv) the format and content of the GMS RGC along with an indication of the gaps that still need to be addressed before regulatory authorities can consider how to develop enforcement measures. The final two sections explain the RPTCC plans for the implementation of the GMS power trading using the RGC (Section 5) and present a summary, conclusions, and recommendations designed to ease the ongoing development by working more closely with stakeholders in power trading (Section 6).

2. GMS Networks and Stakeholders

GMS Transmission Networks

Most HVAC power transmission networks in the GMS extend within their respective national borders throughout an area of 2.5 million square kilometers to serve a population of about 300 million.[7]

As shown in the Map, the main city and urban electricity demand centers are 500 to 1,500 kilometers (km) apart and scattered throughout the region. They are supplied by electricity from hydro, coal and gas. Cross-border grid-to-grid transmission interconnections are also likely to be 100 to 1,500 km long and expected to carry power transfers exceeding 1,000 MW. This will normally require 500 kV HVAC or HVDC overhead transmission lines to ensure stable voltage conditions under full load operations. The topographic map also indicates that routes for transmission interconnections between GMS countries are characterized as mountainous in southern region of the PRC, throughout the Lao PDR, and along the border between Myanmar and Thailand. Overhead transmission lines are likely to traverse thinly populated regions with limited road access, some of which will pass through rainforests and sensitive wildlife habitats. They will also require wayleaves through populated farmland of the Lower Mekong Delta, including areas subject to flooding and with swamps and marshes and areas where landowners will seek significant compensation.

There are significant differences in the capacity of the existing GMS transmission networks and the availability of the generation capacity for power trading. A detailed table of main features of the respective national power development plans is in Appendix 6.[8] It describes the key features of the existing transmission networks along with the development plans and key challenges for each country. Table 2 summarizes the physical characteristics of the existing transmission networks (220 kV, 500 kV HVDC) in the GMS.

The China Southern Power Grid (CSG) is by far the most extensive in the GMS. It has several 500 kV HVAC and HVDC transmission lines connecting the east and west of the CSG, transferring electricity from Yunnan, Guizhou, and Guangxi in the west where the

---

[7]   In comparison with the GMS, the European Union transmission networks within the European Network of Transmission System Operators (ENTSO-e) jurisdiction have similar population densities but in an area of 10 million square kilometers serving a population of about 750 million.

[8]   S. Thorncraft. 2019. *Greater Mekong Subregion Power Market Development: All Business Cases, Including the Integrated GMS Case.* Ricardo Energy & Environment: World Bank.

Greater Mekong Subregion Regional Grid Code
and ASEAN Regional Characteristics

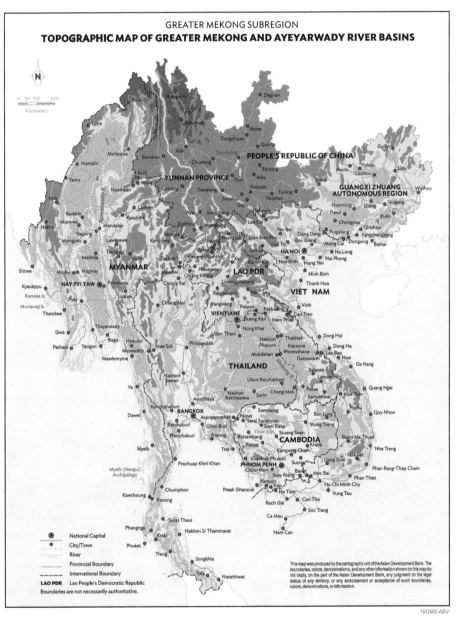

Source: Asian Development Bank.

main generation is located, to Guangdong, the main load center of the CSG. Recently, 800 kV HVDC transmission lines with a rated capacity of 5,000 MW have been built to connect Yunnan and Guangdong. The Yunnan subgrid borders Myanmar and Viet Nam and the Guangxi subgrid borders Viet Nam. Guangxi subgrid is in the center of the CSG and plays an important role in supporting the west–east CSG transmission channel.

Table 2: Physical Characteristics of GMS Transmission Networks in 2020

GMS 2020	PRC (CSG)	Thailand	Viet Nam	Lao PDR	Myanmar	Cambodia
Maximum Demand (GW)	160	35	38	2	5	2
Generating Capacity GW)	323	50	57	4	6	3
Hydro Capacity (GW)		0	17	8	25	4
VRE Capacity (GW)	30	9	12		4	1
Area (km²)	236+394	513	331	236	656	181
Transmission cct km						
500 kV	46,000	6,600	8,000		110	
220 kV	80,000	15,000	17,000	850		1,000
HVDC	7,200	23				
Grid Substations by Voltage Level						
500 kV	125	31	53	16	2	2
220 kV	976	109	8	35	71	28
Cross-Border Lines for Exports	4	1	1	14	2	
Planned Cross-Border Interconnections	3			14		1

GW = gigawatt, km = kilometer, cct km = circuit kilometer, km² = square kilometer, kV = kilovolt, HVDC = high voltage direct current, VRE = variable renewable energy generation (wind or solar power plants).

Source: S. Thorncraft. 2019. *Greater Mekong Subregion Power Market Development: All Business Cases, Including the Integrated GMS Case.* Ricardo Energy & Environment: World Bank.

Of the five ASEAN members in the GMS, Thailand and Viet Nam have the most extensive 500 kV and 220 kV networks that tap into coal and hydropower resources in neighboring Lao PDR. Currently, there are no direct transmission links between Thailand and Viet Nam (i.e., via the Lao PDR or Cambodia) that could be used as a basis for synchronizing the two 500 kV grids. Since the two main demand centers in northern and southern Viet Nam are about 1,300 km apart, there would need to be at least two 500 kV interconnections between the Thailand and Viet Nam grids to minimize problems with separation and resynchronizing after a power system interruption.

The transmission networks in Cambodia, the Lao PDR, and Myanmar were at an early stage of development, with their primary focus on meeting growing domestic and rural electrification programs. Their power systems mostly consist of subtransmission (100 to 115 kV) and medium voltage (20 to 35 kV) distribution networks. The bulk of the Cambodian system is synchronized with Viet Nam, which controls the frequency of the system. A small portion of the 115 kV network is synchronized with Thailand. The Lao PDR system is structured as a domestic and three separate international networks each synchronously connected to the CSG, Thailand, and Viet Nam.

To boost their energy exports, Cambodia and the Lao PDR are planning new enclave power projects with dedicated HVAC radial lines synchronized to the purchasing countries. The Lao PDR is the only country with contiguous borders to all its GMS neighbors and therefore a prime candidate for a common 500 kV or HVDC backbone grid facility linking all six GMS countries. It would also be able to host a secure communications hub that is directly connectable to each TSO via dedicated cross-border optic fiber links.

Expansion of GMS Grid into Adjacent Regions

The GMS has contiguous borders with other southern Asia regions that are also planning to develop power trading among their member countries. In particular, Myanmar shares a border with Bangladesh, and PRC shares a border with India. In both cases, there could be opportunities for the GMS countries to wheel surplus power to the South Asian Association for Regional Cooperation (SAARC). The ASEAN member states also have had a long-standing goal of integrating their power systems by developing the ASEAN Power Grid comprising a series of cross-border HVAC and HVDC interconnectors as shown in Figure 1.

Thailand already has a 300 MW HVDC back-to-back link with the Malaysia grid that is synchronously connected with the Singapore grid. Malaysia is also planning HVDC submarine cable links to Sumatra, Indonesia, and the Malaysian states of Northern Borneo.

Figure 1: Indicative GMS Interconnections for Expansion into ASEAN

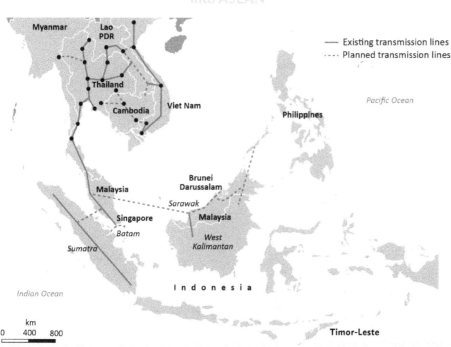

ASEAN = Association of Southeast Asian Nations, GMS = Greater Mekong Subregion, Lao PDR = Lao People's Democratic Republic.

Source. Adapted from International Energy Agency. 2019. *Establishing Multilateral Power Trade in ASEAN*. Figure 5. p. 17.

The PRC has by far the largest integrated hybrid HVDC/HVAC regional power grid in the world. The two synchronously connected grids in Yunnan Province and Guangxi Zhuang Autonomous Region are interconnected by HVDC back-to-back terminals to the much larger northern grid operated by the State Grid Corporation of China and can draw on its significant generation surpluses. The PRC has pioneered new applications for HVDC using voltage source control (VSC) technology with multi-terminal HVDC systems capable of black starting and regulating the frequency of the remote synchronous system. The PRC plans to build several super-high voltages (i.e., 800 kV or above), high-capacity, long-distance direct current transmission projects in parallel with its HVAC transmission networks. Such hybrid HVDC/HVAC networks are expected to be used for power trade with the PRC's northeast neighbors including Mongolia, the Republic of Korea, the Russian Federation, and, possibly, as well as the Republic of Korea and Japan.[9]

Since the PRC began its "going global strategy," banks and state-owned enterprises have made significant investments in regional energy sector projects, whether thermal or hydropower, including within the GMS. In terms of absolute numbers, the PRC is the global leader in the deployment of clean energy technologies, from solar photovoltaic (PV) and wind power to nuclear energy. PRC-based companies are expected to contribute to 15% (54 GW) of power generation development in Southeast Asia between 2013 and 2022.[10]

GMS Interconnection Planning

As shown in Figure 2, the GMS region has a variety of existing HVAC radial cross-border connections for bilateral power trading projects under PPA contracts.

All the independent power producer (IPP) transmission interconnections are radial and mostly HVAC lines originating from hydro and coal powered plants in the Lao PDR. Some of the Lao PDR interconnections are classed as grid-to-grid even though, in effect, they are bilateral enclave projects emanating from one of the four separate synchronous transmission zones within the country. In total, 11 interconnections use 100–115 kV lines, 7 use 230 kV lines, and 6 (including 2 committed lines) use 500 kV. The only HVDC connection is between Myanmar and the PRC. There are 34 radial distribution lines supplying small towns at the borders of the countries involved. A summary of the existing and committed interconnections is in Table 3.

The ADB-funded MHI Master Plan study (Appendix 4) examined the feasibility of the 18 proposed new interconnections ranging from 150 to 1,300 km and capable of transferring 300–3,000 MW on respective lines in stages up to 2030. The study also considered several new interconnection options under a cross-border transmission development scenario. In total, 36 regional generation and transmission planning scenarios were evaluated for power trading from 2022 to 2035 under high-,

---

[9]   Asia Pacific Energy Research Centre. 2015. *Electric Power Grid Interconnections in Northeast Asia.*

[10]   IEA. 2019. *Chinese Companies Energy Activities in Emerging Asia.* April.

Figure 2: Existing and Proposed Cross-Border Transmission
Interconnections

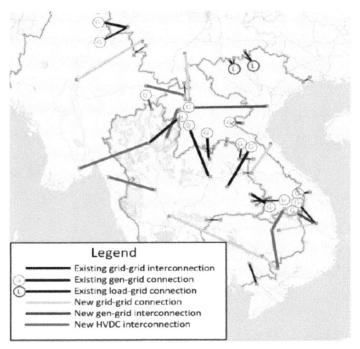

HVDC = high voltage direct current.

Source: Manitoba Hydro International. 2020. Regional Power Master Plan: Harmonizing the Greater Mekong Subregion (GMS) Power System to Facilitate Regional Power Trade. RPTCC 26 Interim Report. October.

medium-, and low-demand growth conditions. Taken together, the planning data and assumptions were used to model national generation and transmission expansion scenarios for development. The study provided outcomes for the scenarios of demand growth in terms of the assumed economic and technological factors with the transmission optimization process ensuring the transmission reliability in system-intact operation (N-0).

The MHI Master Plan medium demand scenario was analyzed in detail to verify the reliability of a cross-border transmission network for an N-1 transmission outage condition. The objective of this analysis was to identify potential transmission congestion situations and propose mitigation options. The MHI Master Plan indicates that the use of synchronous cross-border interconnections under the medium demand scenario would result in an 8–12 GW reduction of new thermal generation development up to 2035. It showed that the development of large-scale coal and gas plants in Thailand and Viet Nam could be avoided or delayed with the use of cross-border interconnections. In contrast, hydro generation in the Lao PDR and Myanmar would increase after enabling cross-border interconnection optimization. Most proposed HVAC interconnections would be designed for upgrading to 500 kV, assuming they will be synchronously interconnected across the region. Section 3 describes some of the technical issues that will need to be considered in effecting interconnections indicating that in the future the GMS is likely to emerge with a hybrid combination of HVAC and HVDC systems.

Table 3: Existing and Committed GMS Cross-Border Interconnectors

| Bilateral Power Exports | | MVAC | HVAC Transmission | | | HVDC | MW.km Ranges | |
From	To		110-115 kV	220 kV	500 kV	+ 220 kV	MW	Km
PRC	Viet Nam			2			300, 300	56, 51
PRC	Viet Nam		3				25-115	20-60
PRC	Lao PDR		1				60	35
PRC	Myanmar		1				75	150
Myanmar	PRC				1	1	240, 600	120 120
Lao PDR	Thailand			2	5		126-1,878	200-350
Lao PDR	Thailand		5				80-160	6-60
Lao PDR	Cambodia	2						
Lao PDR	Viet Nam			2			250, 300	115 120
Lao PDR	Viet Nam	6						
Viet Nam	Cambodia			1			200	111
Viet Nam	Cambodia	18						
Thailand	Cambodia		1				80	40
Thailand	Cambodia	8						
Total Connections		34	11	7	6	1	Total Trans	25

GMS = Greater Mekong Subregion, HVAC = high voltage alternating current, HVDC = high voltage direct current, km = kilometer, kV = kilovolt, Lao PDR = Lao People's Democratic Republic, MV = medium voltage, MVAC = medium voltage alternating current, MW = megawatt, PRC = People's Republic of China.

Source: ADB. 2020. *Regional Power Master Plan: Harmonizing the Greater Mekong Sub-Region (GMS) Power System to Facilitate Regional Power Trade.* Consultant's report. Manila.

The MHI Master Plan recommended 14 specific projects with the largest benefits in priority order indicated by color in Table 4 that also gives cost indicators that can be used as a basis of comparison between each project.

Synchronizing GMS Networks

The MHI Master Plan recognizes that synchronization is a necessary part of its development strategy but is not specific as to how this might be achieved. The report notes that, unless the compatibility issues addressed in the RGC are enforced, there could be system instability problems due to the different dynamic performance of the interconnecting grids. TSOs have also expressed a concern that increased cross-border power trading may create security risks, such as exposure to external system shocks, sudden outages or increased reliance on external resources to meet domestic system needs. However, there have been no detailed technical power system stability studies done to identify the generators that can contribute to primary or secondary frequency control or other system equipment requiring specific protection against overloading.

Table 4: Summary of Recommended Interconnections in the MHI Master Plan

Year	From	To	Connection Points	HVAC Voltage (kV)	Capacity (MV)	Length (km)	Investment Cost ($ million of commission year)	Cost Indicators	
								M$/km	k$/MW/km
2024	Myanmar	Thailand	Yangon area–Mae Moh	500	1,500	350	207	0.59	0.39
			Mawlamyine–Tha Tako	500	1,500	300	184	0.61	0.41
	Lao PDR	Viet Nam	Ban Soc/Ban Hatxan–Tay Ninh via Stung Treng	500	1,000	320	238	0.74	0.74
	Lao PDR	Myanmar	Namo–Kenglat–Tachileik–Kengtung	230	800	230	250	1.09	1.36
	Lao PDR	Viet Nam	Luang Prabang HPP–Sam Nua (Lao PDR–N) Nho Quan	500	2,500	400	432	1.08	0.43
2026	Thailand	Cambodia	Wangnoi–Banteay Mean Chey–Siem Reap Kampong Cham	500	300	500	524	1.05	3.49
	Myanmar	Thailand	Mae Knott TPP–Mae Chan	230	370	115	84	0.73	1.97
	Lao PDR	PRC	Luan Prabang–Yunnan	500	650	350	207	0.59	0.91
	Myanmar	PRC	Mandalay–Yunnan	500	600	350	207	0.59	0.99
2026	Cambodia	Viet Nam	Kampong Cham–Tay Ninh	500	300	100	154	1.54	5.13
2028	Cambodia	Viet Nam	Lower Se San 2 HPP–Pleiku	230	200	230	247	1.07	5.37
2030	Lao PDR	Viet Nam	Xekaman 4 HPP–Ban Soc/Ban Hatxan–Pleiku	500	1,000	120	215	1.79	1.79
			Savannakhet–Ha Tinh	500	600	200	137	0.69	1.14
2031	Lao PDR	Viet Nam	Nam Mo HPP–Ban Ve	220	120	200	219	1.1	9.13
Total					11,440	3,765	3,305	0.88	1.07

HVAC = high voltage alternating current, k$/MW/km = housand US dollar per kilometer and per megawatt, km = kilometer, kV = kilovolt, Lao PDR = Lao People's Democratic Republic, M$/km = million US dollar per kilometer, MHI = Manitoba Hydro International, MV = medium voltage, MW = megawatt, PRC = People's Republic of China.

Source: ADB. 2020. *Regional Power Master Plan: Harmonizing the Greater Mekong Sub-Region (GMS) Power System to Facilitate Regional Power Trade*. Consultant's report. Manila.

There have been two recent reports with conceptual plans for synchronizing the GMS countries in a staged process. In contrast, the 2020 Japan International Cooperation Agency (JICA) technical study of the Lao PDR national transmission plan proposes a more conservative approach to synchronization, with the Lao PDR grid interconnecting with Cambodia and Thailand by 2040 through a 500 kV network. The JICA report recommends that the existing IPPs in the Lao PDR should continue to operate as lateral synchronous suppliers to the Thai network to avoid problems with circulating power flows. Although the report does not support the contention with system studies, it considers this situation likely if the IPPs and the Lao PDR 220 kV transmission networks were all interconnected synchronously through the Lao PDR.

The 2019 World Bank report (Appendix 4) targeting 2030 for full synchronization is shown in Figure 3. It has the most detailed proposal for how this could be achieved in four stages of development.

There is no common agreement on a future backbone structure for a 500 kV or HVDC GMS grid with competing options being promoted by different interests. There is, however, an urgent need to undertake comprehensive load flow, fault, and stability studies to identify measures that can be taken to mitigate any instability of the respective networks if they are synchronized. On the other hand, the initial deployment of HVDC interconnections would not require rigid compliance with the

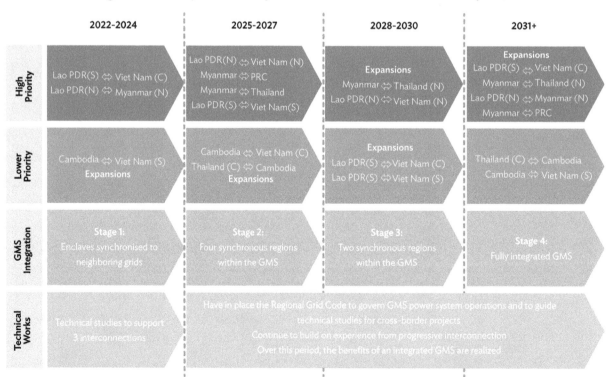

Figure 3: Concept Plan for Synchronization of GMS Network by 2030

C = central, GMS = Greater Mekong Subregion, Lao PDR = Lao People's Democratic Republic, N = central, PRC = People's Republic of China, S = south.

Source: World Bank. 2019. Greater Mekong Subregion Power Market Development: All Business Cases including the Integrated GMS Case. Ricardo.

RGC and could be used to help build confidence in effecting GMS regional power trading while generator control system upgrades are being made. This is because HVDC power flows can be closely controlled by TSOs, while the terminal converters can also help maintain system stability across the two interconnected grids.

EDL-T Transmission Company

The Électricité du Laos Transmission Company Limited (EDL-T) was established as a joint venture between the CSG and EDL-T. The Concession Agreement signed on 11 March 2021 enables EDL-T to serve as the country's national power grid operator under the supervision of the Government of the Lao PDR. EDL-T will invest, construct, and operate power grids (230 kV and above), and implement grid interconnection projects between the Lao PDR and its neighbors. The new projects will also include a new National Control Center expected to be in operation by 2022.

According to the concept plan presented in the JICA report, an objective of the joint venture is to effect a 500 kV synchronous interconnection between the PRC and the northern and central Lao PDR. This would be connected asynchronously through HVDC connections with Cambodia and Viet Nam and achieved with HVDC back-to-back converter stations located both in the north and south of Viet Nam and direct HVDC lines to Cambodia in the south. If implemented, the EDL-T proposal could have a significant impact on the recommendations outlined in the MHI Master Plan, although the project is yet to be discussed within the RPTCC. A concept plan for a possible hybrid 500 kV/HVDC grid under consideration is shown in Figure 4.

Stakeholders in the GMS Power Trade

Aside from the GMS TSOs who developed the RGC, there are many other stakeholders whose interests in power trading would be impacted by its being enforced to govern the proposed interconnections and power trading within the GMS. They include owners and operators of generation assets, TSO and distribution system operators (DSOs), electrical manufacturers and their associated installation contractors, along with directly connected large industries. Ultimately, consumers would want to see the benefits of regulated power trading in terms of higher reliability, lower costs, and reduced emissions. To ensure stakeholders' acceptance of the RGC, it is necessary they are all involved in its further development and implementation.

There are also several international financial institutions, bilateral aid organizations, and other development partners supporting the energy sector in the GMS, along with other representative bodies such as the members of the ASEAN Power Grid. The main development partners have produced or funded studies on the energy resources and technologies available to determine the opportunities for power trading within the GMS member countries. Other studies have included reviews of the existing NGCs and the scope for their development to foster power trade. In its capacity as regional coordinator, ADB has managed most of the bilateral grants to the GMS countries and plans to continue fostering cross-border power trade within the GMS.

Figure 4: Proposed EDL-T Plan for the HVDC/HVAC Hybrid Grid
in the Lao PDR by 2030

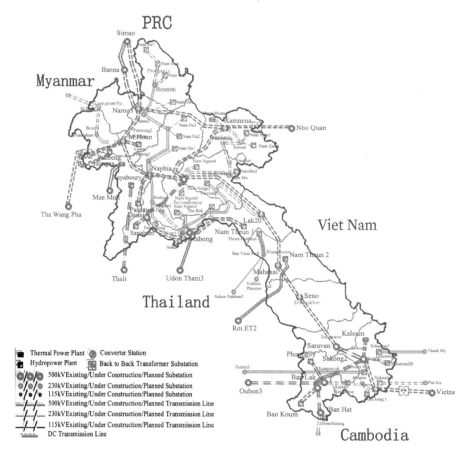

DC = direct current, EDL-T = Électricité du Laos Transmission Company Limited, HVAC = high voltage alternating current, HVDC = high voltage direct current, kV = kilovolt, Lao PDR = Lao People's Democratic Republic, PRC = People's Republic of China.

Source: JICA. February 2020. Power Network System Master Plan in Lao PDR, Chapter 7.

GMS National Regulators

A summary of the status and organizational responsibilities of the GMS national institutional and regulatory frameworks is described in Section 5. In brief, the three largest GMS grids (CSG, Thailand, and Viet Nam) have functioning regulatory systems. CSG's regulator, the State Electricity Regulatory Commission, was established in 2002 and Cambodia's Electricity Authority of Cambodia was established in 2001. The Electricity Regulatory Authority of Vietnam was set up in 2005 and Thailand's Energy Regulatory Commission in 2008. Only Thailand is a member of the Energy Regulators Regional Association (ERRA) whose mission is to support the strengthening and improvement of the regulatory framework and being a reference point for practice and knowledge in the constantly evolving regulatory environment.[11]

---

11   The ERRA website provides a list of 50 members and associates. https://erranet.org/about-us/members/.

Among the GMS countries, NGCs are available in the PRC (2011), Cambodia (2009), the Lao PDR (2008), Thailand (2012), and Viet Nam (2010–2019). The PRC grid code is believed to be comprehensive and to generally follow ENTSO-e standards but is not available in English.[12] Otherwise, many NGCs have been developed on an ad hoc basis as new issues were identified during operations. As noted in Section 5, this has been confirmed in a review of the grid codes for Cambodia, the Lao PDR, and Myanmar where there were many omissions and ambiguities that needed to be addressed as the new GMS RGC was being implemented.

A recent review that compared ASEAN grid codes concluded that Thailand had the most comprehensive coverage of grid codes developed continuously by the TSO (Energy Generating Authority of Thailand [EGAT]), along with the DSOs, the Metropolitan Electricity Authority and the Provincial Electricity Authority under supervision of the regulator of the Energy Regulatory Commission, as presented in Figure 6.[13] The review concluded, however, that the EGAT transmission grid code is aimed at conventional generation and does not address the characteristics of intermittency and low inertia inherent in VRE installations. The two DSOs have been separated under the aegis of the Energy Regulatory Commission of Thailand so do not always align with the EGAT NGC. There is some communication on requirements between EGAT and the two distribution operators, as EGAT requirements are mentioned in the Provincial Electricity Authority code, and some requirements for general power system stability are included in the distribution codes. No information about stakeholder involvement are provided in any of the distribution codes. This is probably an appropriate time to consider bringing all three codes into one common "whole-of-system code" as part of the process of aligning them with the new GMS RGC.

A similar review of the Viet Nam grid code also identified weaknesses in the existing rules regarding the variety of policies for system security, quality of service, and smart grid policies that Viet Nam has since addressed.[14] More recently, a much more detailed review of the Viet Nam grid has made specific recommendations to consolidate the codes into one document, which would make the respective TSO/DSO codes more attractive to investors in renewables.[15] They will be designed to ensure that operational aspects are in place, and that information exchanged are valid and present at the requested time and of the required quality and include the missing connection codes for demand, storage, and HVDC facilities. They will also address the market code and especially the ancillary services, although these are not considered critical for the security of supply in the short term and midterm.

---

[12]   D. W. Gao et al. 2016. *Comparison of Standards and Technical Requirements of Grid-Connected Wind Power Plants in China and the United States.* National Renewable Energy Lab (NREL). 1 September. https://doi.org/10.2172/1326717.

[13]   ASEAN Centre for Energy (ACE), GIZ, T. Ackermann, E. Troester, and P.-P. Schierhorn, eds. 2018. *Report on ASEAN Grid Code Comparison Review.* Jakarta. October. https://aseanenergy.org/report-on-asean-grid-code-comparison-review/.

[14]   World Bank. 2016. *Smart Grid to Enhance Power Transmission in Vietnam.* Washington, DC. https://openknowledge.worldbank.org/handle/10986/24027.

[15]   Energinet Associated Activities. 2020. *Grid Codes: Comparison of Vietnamese and European Grid Codes.* October. https://ens.dk/sites/ens.dk/files/Globalcooperation/grid_codes_d31_oct_2020_2.pdf.

Within the GMS, each country is undergoing some form of restructuring with the aim of developing an open energy market with multiple active entities requiring clear and transparent guidelines and communication between market players. This stage of grid-to-grid power trading is seen as an opportunity to enable the GMS countries to build upon some elements established in the bilateral model, such as introducing harmonized wheeling charges and new elements such as a regional market operator and a central clearing function. To enable the RGC to be incorporated into the laws of each country, the respective national regulators must be given the opportunity to review and participate in its development. Official recognition of the GMS RGC by national regulators will help modernize energy sector regulation to establish a more effective electricity market system, eventually unbundling integrated electric grid operators and reducing electricity costs to consumers. These concerns about the efficacy of the existing GMS NGCs are expected to be reviewed under the recently agreed Japan–United States Mekong Power Partnership.[16]

GMS Power Market Entities

Originally, the PRC power system was characterized as a centrally planned organization. With gradual sector reform, the current system combines both centrally planned and market-based elements and remains in a process of transformation.[17] Responsibility for power dispatch has been assumed by provincial dispatching centers affiliated with grid companies and have a multi-level hierarchy. Provincial dispatching plays the most important role in the system, while national and regional dispatch centers take charge of interregional and interprovincial power transmission, and prefectural dispatch centers are mainly responsible for prefectural load management.

Thailand and Viet Nam are in the process of restructuring to reduce the cost of power and meet their global obligations to reduce carbon emissions by increasing competition and improving access to power trading. As shown in Figure 5, the TSO in Thailand (EGAT) operates as a single buyer market and sells its power to two main DSOs (the Municipal Electricity Authority [MEA] of Bangkok and the Provincial Electricity Authority [PEA] of Thailand) and large industries. Viet Nam established a domestic wholesale electricity market in 2019 and intends to establish a retail market by 2024. However, regulatory consideration will be needed to establish how international power trade will be integrated into the operation of the Viet Nam market, possibly requiring a special carve-out arrangement to accept imported grid-to-grid power in its balancing regime.

The power market entities in Cambodia and Myanmar were by and large vertically integrated organization structures incorporated within their respective energy ministries. It is expected that as these countries become more involved in power trading, they are likely to evolve structures somewhat like Thailand's. The situation in

---

[16]   United States Embassy and Consulate in Vietnam. 2020. Mekong-U.S. Partnership Joint Ministerial Statement. 15 September. https://vn.usembassy.gov/mekong-u-s-partnership-joint-ministerial-statement/.

[17]   International Energy Agency. 2019. China Power System Transformation: Assessing the Benefit of Optimized Operations and Advanced Flexibility Options. February. https://www.iea.org/reports/china-power-system-transformation.

the Lao PDR is more complex, with a clear separation between the country's smaller domestic market and its rapidly growing and much larger export market (Figure 6). Interconnections between GMS stakeholder trading entities are likely to be based on mutually agreed grid codes like those developed in Thailand between EGAT, MEA, and PEA and regulated by the Energy Regulatory Commission of Thailand as indicated in the relationships shown in Figure 6.[18]

In contrast, the power sector organization structure in the Lao PDR will involve separate entries for domestic supplies and international supplies and has a diverse group of stakeholders that will also have an interest in power trading issues (Figure 6).

Distribution Network Operators

Distribution network owners are likely to participate as stakeholders in regional power trade, especially as they develop distributed generation resources throughout their networks. For example, in Thailand, the PEA already has a significant share of renewables connected to its networks and may well be interested in wheeling through EGAT grid surplus photovoltaic (PV) solar power to MEA on hot days when the air-conditioning demand is high. While distribution network operators have long-established codes of practice for interconnections at substations, they were essentially a form of one-way grid-to-grid connections. The connections are normally made where the transmission voltage (typically 220/500 kV) is reduced to a lower voltage (typically 20–100 kV) to enable the distribution network operator to expand its networks safely in constrained urban or rural areas.

Figure 5: Organizational Structure of Thailand's Power Trading Stakeholders

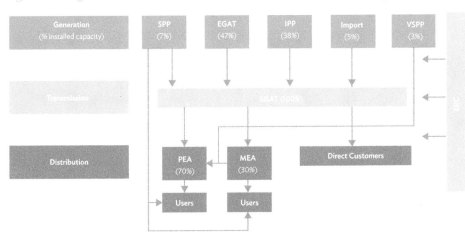

EGAT = Electricity Generating Authority of Thailand, ERC = Energy Regulatory Commission of Thailand, IPP = independent power producer, MEA = Metropolitan Electricity Authority, PEA = Provincial Electricity Authority, SPP = small power producer, VSPP = very small power producer.

Source: ASEAN Centre for Energy (ACE), GIZ, T. Ackermann, E. Troester, and P.-P. Schierhorn, eds. 2018. *Report on ASEAN Grid Code Comparison Review.* Jakarta. October. Figure 5, page 36.

---

18   All three utilities publish their own grid codes. The main applicable documents are the EGAT Grid Code of 2014, the PEA Interconnection Code of 2016, and the MEA Interconnection Code of 2015.

Very soon, TSO/DSO power is likely to flow in both directions as distributors ramp up their support for solar power, grid battery installation, and vehicle-to-grid schemes.[19] There is also increasing interest within DSOs in offering aggregated demand-side management schemes using the potential storage capacity inherent in heating and cooling systems that can be switched on and off to manage peak loadings on the network. There is a new initiative in the United Kingdom to review the transmission and distribution codes with a view of their being simplified in the form of a digitized whole-of-system grid code.[20]

Independent Power Producers

Within the GMS countries, there has been a boom in private sector power trading by construction companies, hydropower developers, and private banks. Although most of these IPP stakeholders are from Asia (within and outside the GMS) there are others from outside the region as well. Most privately owned IPPs have invested in long-term hydro generation and associated transmission projects but generally not in TSO or DSO network infrastructure. Moreover, the potential generation capability in these situations is increasingly getting locked into long-term PPAs that may not include adequate technical clauses for voltage and frequency controls to facilitate grid-to-grid power trading. Any future power trading regulatory initiatives, including enforcement of RGC requirements, will need to address IPP concerns about potential changes to their contracting arrangements.

Figure 6: Organizational Structure and Stakeholders in the Lao PDR

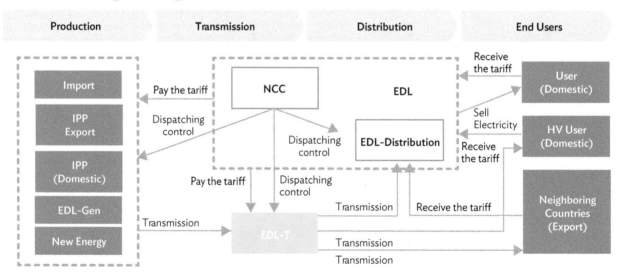

EDL = Électricité du Laos, EDL-T = Electricite du Laos Transmission Company Limited, HV = high voltage, IPP = independent power producer, NCC = National Control Center.

Source: Japan International Cooperation Agency. 2020. *The Study on Power Network System Master Plan in Lao People's Democratic Republic*: Final Report. February. Chapter 7.

---

[19]   L. Jones et al. 2021. The A to Z of V2G: A Comprehensive Analysis of Vehicle--to--Grid Technology Worldwide. Australian Renewable Energy Agency Battery Storage and Grid Integration Program. January. https://arena.gov.au/assets/2021/01/revs-the-a-to-z-of-v2g.pdf.

The increased use of hydropower and pumped storage is on the GMS agenda for large-scale storage but the commercial viability of this option is country-dependent. In this respect, IPP ownership of large hydropower reservoirs in the Lao PDR and Myanmar could have a leading role in managing the growing need for flexibility in the power grid. Their storage reservoirs can provide large quantities of both capacity (short-term power flexibility) and energy (medium-term and long-term power and energy flexibility). In recent developments, there have been several floating PV installations that can be used in conjunction with the storage capability of the associated hydro plant to enhance its production profile.

Directly Connected Industries

In the GMS, there are many industrial consumers with large electricity loads that connect directly to GMS networks. These can have significant adverse impacts on grid stability and the quality of supplies that, without mitigation, can be transmitted across national boundaries. These organizations use power electronics comparable to generators and thus impact the stability of the grid system in a similar manner. Electric arc furnaces are often one of the largest loads in power systems that are highly nonlinear and time-varying which cause power quality problems such as harmonics and flicker. Some industries require uninterruptable power and may need to be supplied from independent sources. There is also the potential for large industries to participate in demand-side management opportunities to make changes to their production processes to curtail their electricity usage during peak periods. Other industries may have generation capacity feeding into the interconnection with transmission and will, of course, be subject to the same obligations as any other generating plant.

Electrical Equipment Manufacturers

Manufacturers of electro-mechanical equipment that make up terminal substations, transformers, switchgear, metering equipment, flexible alternating current transmission systems (FACTS), or converter facilities must meet stringent international standards designed to ensure that competing suppliers conform to the same technical conditions. A significant number of manufacturers and installers of switchgear, metering and control systems, transformers, and transmission lines are based in both the PRC and Southeast Asia. Some GMS companies are making such equipment for export, often under license to larger international companies. With the advent of better-performing materials and more sophisticated control systems, even these standards are increasingly coming under the umbrella of the grid codes. Accordingly, when new RGCs are being developed for the GMS, such that new equipment designs or construction techniques may be required, it is important for manufacturers and installers to be stakeholders in the deliberations.

---

[20]   National Grid ESO. 2021. Digitalized Whole System Technical Code Consultation. https://www. nationalgrideso.com/document/197521/download.

3. GMS Interconnection Options

Planning to Facilitate the GMS Synchronization

Feasibility Studies

The MHI Master Plan studies are largely focused on the expansion of interconnections in anticipation that the ASEAN group of GMS countries will quickly achieve a fully synchronized system. However, in comparison with the European Union (EU), where regional synchronization was established over 40 years ago without an RGC, it is proving to be a challenge for the GMS to take its first steps toward this goal. Because it will be some time before GMS RGC requirements for harmonization of generation and power system operations can be enforced, the process of synchronization in stages will need to be carefully managed. However, as more grid-to-grid interconnections are made, there should be a general increase in the security, reliability, and stability of the larger GMS power system. Common planning standards will be required to complement the RGC by setting limits on transmission system voltage and frequency variations, fault events, reactive power capabilities, safety, and security. NGCs will also need to be continuously updated to cope with the integration of renewables along with their mitigating technologies to ensure reliability and power system stability.

The best way to consider the technical issues of interconnections is to group them into those associated with the transmission interconnection itself, and those associated with operating the larger interconnected system. Those associated with operating the larger interconnected system will require the sharing of national transmission planning data to carry out comprehensive load flow, fault, and system stability studies of the interconnected grids, taking into account different contingency situations with critical elements of the power systems in or out of service. The main issues to consider include thermal limits, stability limits, and voltage regulation, which are the main constraints on transmission line operation. Other transmission issues include loop and parallel path flows, available transfer capacity, and the capability of FACTS technologies to increase capacity transfers by mitigating voltage instability issues. The main criteria for evaluating the consequences of interconnections between grids should consider the following:

(i) **Cascade tripping**. A single contingency event should not result in any cascade tripping in the respective grids.
(ii) **Thermal limits (normal and overload ratings)**. Normal operations and all single contingencies must not result in a permanent excess of the permissible rated limits of the network equipment.

(iii) **Voltages**. Normal operations and all single contingencies must not result in permanent violation of the permissible voltage limits on the busbars of all transmission system substations.

(iv) **Loss of demand or generation capacity**. Power available in primary regulation reserves for each synchronous region should not be exceeded.

(v) **Short circuit levels**. The short circuit current rating of the equipment should not be exceeded.

(vi) **Stability conditions**. Three-phase short circuits with subsequent fault clearing in any of the elements of the transmission system must not result in the loss of synchronization for any generating unit unless the short circuit is located directly in the power plant.

(vii) **Angular difference**. The angular difference limits should not be exceeded to ensure the ability of circuit breakers for reclosing, without imposing unacceptable step changes on local generation.

Protection systems are an extremely important part of the power networks and must also be shown to be compliant with the RGC as part of the analytical studies. Their primary function is to detect and clear faults, which are inadvertent electrical connections, that is, short circuits between system components at different voltages. When faults occur, extremely high currents can result, typically 2 to 10 times as high as normal load currents. Since power is proportional to the square of the current, a great deal of energy can be delivered to unintended recipients in a short time. The goal of protection systems is to isolate and de-energize faults before they can harm personnel or cause serious damage to equipment. Protection systems must be designed to protect the power system itself, rather than end-user equipment.

Design Standards

It is only through technical designs that meet recognized international manufacturing standards that the requirements of interconnectivity and interoperability between competing products, services, and processes can be assured. Structural and equipment design standards have been promulgated by international contractors and are constantly being updated as new material technologies arrive on the market. As the RGCs become an important part of grid-to-grid operations, both design codes and grid codes will need to adapt to the changing circumstances.

The most important international equipment standards are the ones issued by the International Organization for Standardization (ISO), International Electrotechnical Commission (IEC), Standardization Administration of the PRC (SAC) and the US-based Institute of Electrical and Electronics Engineers (IEEE).[21] Other national standards or codes of practice are also important but they are often adaptations of standards issued by these key organizations. In some cases, national standards are not recognized internationally because they overtly enable local manufacturing capability that does not always meet stringent criteria applied internationally.[22]

---

[21]   D. Narang et al. 2020. *Clause-by-Clause Summary of Requirements in IEEE Standard 1547-2018*. NREL. March. https://www.nrel.gov/docs/fy20osti/75184.pdf.

[22]   The power industry has traditionally applied standards used by major manufacturing countries particularly those recognized by international procurement agencies including British Standards (UK), Deutsches Institut für Normung (Germany), Japanese Industrial Standards (Japan), and Standardization Administration of China (PRC).

In terms of installation and operations, there are codes of practice for safety that are mandatory in most jurisdictions. These cover issues such as wiring codes, electrical safe distances, earthquake standards, and wind loadings. The designer of a typical transmission line must consider a variety of conditions essential to meeting grid code criteria for reliability, safety, and economy of operation. The transmission line would typically be rated to consider environmental conditions such as wind and sun loadings, routing conditions, types and strengths of structures, and distance between grids. The materials used in its construction are required to meet international civil and electro-mechanical standards relating to the environment through which it passes. These would include consideration of ambient and worst-case temperatures, wind, and icing conditions, isokeraunic levels, and foundation design as they might affect service conditions over a typical lifetime of 40 years.

Construction Planning

The MHI Master Plan was developed using standard distance and capacity-based cost rates for HVAC and HVDC lines and terminal connections. Although its investigations included load flow studies to determine if associated grid reinforcement was necessary, the associated costs were also based on indicative rates for mitigation measures. After identifying and recommending priority interconnection candidates for development, the MHI Master Plan recommended that detailed feasibility studies be carried out. These would help determine the design parameters of each transmission project in terms of its routing, terminations, and environmental and protection requirements. The analytical studies would need to evaluate the impact of power trading flows on the respective grids in terms of voltage stability, fault carrying capacity, circulating currents, metering, and protection. This work would need to be followed up with a detailed survey of the line route to get an accurate assessment of the construction conditions including provision for access for maintenance during operation. The survey would normally involve the identification of tower locations to define the sag and span profiles and for conducting environmental and land compensation assessments. Eventually, all this information would be used for detailed project costing to establish the basis for setting tariffs and institutional arrangements as required for project financing.

The GMS cross-border interconnections are expected to be mainly by HVAC or HVDC overhead transmission lines. The technical aspects of each project would need to be designed so they follow the GMS RGC codes, particularly regarding operational security requirements considering the following:

(i) Reinforcement of overhead circuits to increase their capacity (e.g., replacing circuits, increased distance to ground).

(ii) Duplication (bundling) of conductors to increase rating, upgrading of network equipment or reinforcement of substations (e.g., based on short circuit rating).

(iii) Extension of substations and construction of new ones.

(iv) Installation of reactive power compensation equipment (e.g., capacitor banks).

(v)   Addition of network equipment to control the active power flow
       (e.g., phase shifter, series compensation devices, flexible alternating
       current transmission systems).
(vi)   Additional transformer capacities.
(vii) Construction of terminal facilities (overhead and cable).

Overhead Transmission Lines

The basic elements of a double circuit HVAC transmission interconnection include
three conductors per circuit, insulators, support structures, transformers and
substations, protection systems, communications, and monitoring and control
systems. An HVDC transmission line capable of carrying the same load as an HVAC
line would cost about 40% less to construct. However, both ends of an HVDC
transmission interconnection would need to terminate at an HVDC/HVAC converter
facility (typically costing $100 per kilovolt-ampere [kVA] per terminal) before being
connected to a conventional HVAC substation (costing up to $30 per kVA).

Synchronous HVAC Interconnections

HVAC interconnections are the most common for distances less than 400 km and
are usually the cheapest way of effecting synchronous grid-to-grid power trading.
By themselves, HVAC transmission lines do not involve any significant costs other
than the construction of the line and the terminations in the associated substation
switchgear equipment. Normally, two large grids would need to be resynchronized
each time an HVAC interconnection is interrupted, typically caused by common
events such as lightning strikes or bird strikes and often resulting in a chain reaction
and blackouts. Under black start conditions, the weaker system would need to
manage the staged reconnection of its generators to match the other grid's frequency.
Accordingly, to maintain system security, it is necessary to have two or more separate
HVAC circuits connected to keep both system grids synchronized if any one circuit fails.

Transmission transfer capability (TTC) is the maximum power flow that a line
can accommodate at any given time and still be able to survive the loss of a major
generator or transmission link elsewhere in the system. Available transmission
capacity (ATC) is the TTC of a line minus the amount of capacity already committed
to other uses on that line. ATC is thus the measure of how much power can be safely
transmitted over a transmission line at any given time while ensuring overall system
reliability. In this respect, an HVAC line can only transfer power according to its
"loadability," which is a function of the voltage, conductor and bundling sizes, and
the length of the line. For short 500-kilovolt (kV) lines (i.e., up to 100 km), the most
economic loadability will be about half the maximum thermal rating of the line, which
is determined by the aggregate bundled conductor cross-sectional area, typically
about 1,200 MW/circuit. For longer 500-kV lines, the loadability is decided by the
surge impedance loading of the line (i.e., about 500–700 MW), tower structures,
and conductor spacing. For HVAC lines longer than 400 km, stability limits may
decide the largest allowable load, which can be as low as 20% of the thermal rating.

Normally, unless there are plans to install intermediate substations to supply a local load, this would be a crossover point in comparing the cost of long HVAC and HVDC interconnections.

Upgrading HVAC Transmission Capacity

HVAC transmission lines of more than 200 km may need intermediate substations for the installation of FACTS facilities to ensure proper voltage regulation along transmission lines. Typically, FACTS equipment used in transmission grids includes a variety of functions (Figure 7) indicating the range of capabilities in the standard Electric Power Research Institute (EPRI) model.[23] The model would be used in power system studies to evaluate the following services:

(i) **Series compensation**. This technology provides improved grid stability and contributes to the optimal use of transmission lines. Among the main benefits of series compensation are the reduction of line voltage drops, limitation of load-dependent drops, and a reduction of the transmission angle.

(ii) **Static VAR compensation.** This equipment is used to increase grid reliability by assisting fault recovery and thus reduce the risk of blackouts. SVCs can improve the power factor by dynamically providing reactive power. They can symmetrize unbalance between phases, and reduce flicker for large industrial consumers.

(iii) **STACOM.** This equipment is used for reactive power compensation to increase dynamic stability and the power quality of the grid; it is based on multi-level VSC technology.

(iv) **Static VAR compensation frequency stabilizer.** This is used when renewable sources continuously replace conventional synchronous power generation and the grid frequency is becoming more sensitive due to the reduced amount of synchronous generators. It will provide sufficient system inertia to stabilize the grid.

(v) **Synchronous condenser.** A synchronous condenser provides short circuit power, inertia, and reactive power for dynamic loads.

(vi) **Mechanically switched capacitors.** These provide a simple and low-speed solution that provides grid stabilization and voltage control under heavy load conditions, while mechanically switched reactors provide stabilization under low load conditions.

Asynchronous Connections

Variable Frequency Transformer Asynchronous Connections

A variable frequency transformer (VFT) is a relatively new device that can be used to transmit electricity between two alternating current frequency domains (asynchronous or synchronous). Most asynchronous grid inter-ties use HVDC

---

[23]   IEEE/EPRI. 2011. HVDC & FACTS Research at EPRI. https://www.epri.com.

Figure 7: US Electric Power Research Institute Generalized Model
of Flexible Alternating Current Transmission Systems Devices

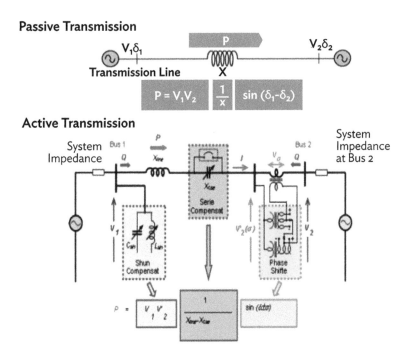

US = United States.

Source: EPRI. 2011. *EPRI HVDC Research* prepared for the EPRI HVDC and Facts Conference.
30 August. http://mydocs.epri.com/docs/publicmeetingmaterials/1108/6XNSUMJE9MT/EPRI_
HVDC_&_FACTS_Research_Program-Conference2011.pdf.

converters, while synchronous grid inter-ties are connected by lines and ordinary
transformers, but without the ability to control power flow between systems.
The technology shown in Figure 8 was pioneered by Quebec Hydro and has been
applied successfully in several North American jurisdictions.

The VFT behaves as a continuously adjustable phase-shifting transformer. It can
be thought of as a very high-power rotary converter acting as a frequency changer,
which is more efficient than a motor generator of the same rate.

VFT is a relatively inexpensive solution based on well-established alternating
current technology with no harmonic generation. VFTs provide the ability to make
power flow in either direction between two grids, permitting previously impossible
power exchanges. VFTs are also used in large land-based wind turbines so that the
turbine rotation speed can vary while connected to an electric power distribution grid.

A VFT facility used in the GMS region, at about 30% of the cost of an HVDC
back-to-back facility, could be suitable for the lower capacity 220 kV
interconnections between the smaller grids, for example, between Cambodia and
Viet Nam or the Lao PDR to facilitate power trading on a controlled basis to help both
sides build confidence in the procedure.

Figure 8: Cutaway View of a 100-Megawatt Variable Frequency Transformer

Source: D. Nadeau. 2007. A 100-MW Variable Frequency Transformer (VFT) on the Hydro-Québec TransÉnergie Network--The Behavior during Disturbance. *Institute of Electrical and Electronics Engineers.* 10.1109/PES.2007.385584.

HVDC Interconnections

There are technical, cost and reliability advantages of interconnections using long, high-capacity HVDC lines. HVDC was originally developed to supply large volumes of power over long distances using line commutate converter (LCC) technology. This system can only function if the two connecting systems are already in operation. The first large-scale commercial project was installed in 1965 in New Zealand where a 600-megawatt HVDC line was built to carry power from the South Island 600 km to the North Island. This inter-island link has operated reliably for over 45 years and was recently upgraded to 1,400 MW. Since 1965, over 200 HVDC systems have been built around the world.

A typical HVDC transmission line comprises two circuits (one positive, one negative) together with a ground return circuit for use when any one conductor suffers a line short circuit or insulation failure. When both HVDC circuits are in full operation, there will be a potential difference of double between them and no current flows through the ground circuit. However, each circuit is capable of 50% of the rating using the ground return if the other circuit fails.

Because direct current transfers only active power and thus cause lower losses than alternating current (which transfers both active and reactive power), HVDC transmission losses are normally quoted as less than 3% per 1,000 km, which are 30% to 40% less than with alternating current lines at the same voltage levels. HVDC can be switched on and off without the complications and delays involved in the re-synchronism procedures of large neighboring power systems. Importantly, HVDC terminals can also act like fast-acting batteries capable of mitigating intermittency of wind or solar generating plants. In this respect, they can be used to earn additional revenue in a power market requiring ancillary services to maintain power system stability.

HVDC systems can also be built in stages to increase loading as required. Figure 9 shows how this can be done in combinations by first building a line for monopole operation, then uprating the converters to bipole operation and, if the transmission line has been appropriately insulated, uprating again using a higher operating voltage. Given that the line can be designed for its ultimate operating configuration (at little extra cost), the cost lies primarily in uprating the HVDC/HVAC terminals at each end of the line.

The popularization of the HVDC-VSC technology is now finding new applications for black starting on systems and providing the necessary inertial power to bring both systems into synchronism. Their HVDC or HVAC converter terminals can provide important stabilizing functions to enhance the security of the receiving HVAC systems by enabling power to be tightly controlled by the TSOs.

Regional interconnections through HVDC-VSC links can effectively improve the transfer capability between regional networks. The precise power flow control of HVDC links simplifies the settlement of pricing power transfers as well as customer billing and prevents free riders. HVDC-VSC systems can also be operated as a merchant transmission facility, like a merchant generator. Another advantage is that the power direction is changed by changing the direction of the current and not by changing the polarity of the HVDC voltage. This makes it easier to build HVDC–VSC systems with multiple terminals connected to different points in the same or different grids. The resulting HVDC grids can be radial, meshed, or a combination of both. At locations where two asynchronous HVAC grids have extensions in the same vicinity, typically at national borders, the two terminal converters could be installed in the same building in HVDC/HVAC in a back-to-back configuration. However, such installations are expensive and likely to become stranded investments after the two interconnecting grids are fully synchronized.

Hybrid HVDC and HVAC Systems

Once GMS grids are fully synchronized HVAC (500 kV and 220 kV), many more interconnections can be expected to be deployed throughout the national transmission, city and rural distribution systems. However, it is likely that HVDC will be the preferred option to begin the process of building grid-to-grid interconnections for power trading, particularly where they can be used to share reserve capacity and serve to stabilize the underlying HVAC systems. There are many examples where an HVDC line is superimposed to operate in parallel with synchronous HVAC systems and more are expected to be developed as multi-terminal systems. The first major project of this type, built in 1970, was the 1,400 km HVDC Pacific Northwest DC Inter-tie that transmits electricity from the 1,400 km Pacific Northwest to Los Angeles bypassing the HVAC interconnected systems serving the individual states enroute. There is also a global trend toward greater use of hybrid HVDC/HVAC systems with the EU now developing a strategic plan to move in this direction.[24]

---

[24]   ENTSO-E. 2019. *Vision on Market Design and System Operation towards 2030.* https://vision2030.entsoe.eu.

Figure 9: HVDC Inverter Monopole and Bipole Configurations

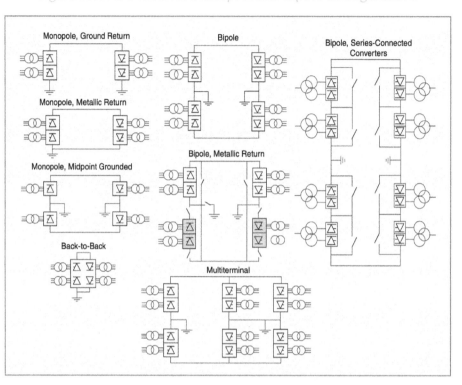

HVDC = high voltage direct current.

Source: M. P. Bahrman and B. K. Johnson. 2007. The ABCs of HVDC Transmission Technologies. *IEEE Power & Energy Magazine.* 5 (2). March–April. pp. 32–44.

Figure 10 shows schematically how a hybrid system using an HVDC and an HVAC/FACTS system would operate. Power exchange in the neighboring areas of interconnected systems offering most advantages can be realized by HVAC links, often including FACTS for increased transmission capacity and for stability reasons. The transmission of larger power exchanges over longer distances would normally use HVDC transmissions directly to the locations of power demand. HVDC can be realized either as a direct coupler without the intermediate transmission line, the back-to-back solution, or as point-to-point long-distance transmission via an interconnecting HVDC transmission line. The long HVDC links can also be used to strengthen the underlying HVAC interconnections, avoiding possible dynamic problems that can exist in the underlying synchronous grids as shown in Figure 10.

Figure 10: Large Power System Interconnection—Benefits
of Hybrid Solution

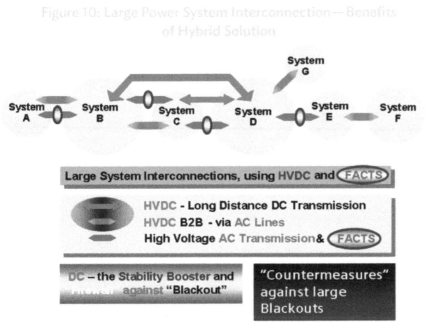

Figure 10: Large Power System Interconnection—Benefits of Hybrid Solution

AC = alternating current, DC = direct current, FACTS = flexible alternating current transmission systems, HVDC = high voltage direct current.

Source: ENTSO-E. 2019. *Vision on Market Design and System Operation towards 2030*. https://vision2030.entsoe.eu.

4. Grid Code Types and Applications

National and Regional Grid Codes

Grid Code Origins

All grid codes are agreed sets of rules, specific to the safe and reliable connection and disconnection of two or more electrical entities designed to enhance the security and economy of the production, transport and consumption of electricity. There are over 400 types of grid codes in 65 countries as listed on the Det Norske Veritas (DNV) website.[25] Mostly, they are designed as NGCs to enable a country's regulators, transmission and distribution operators, generators, suppliers, and consumers to function more effectively across the market. NGCs broadly cover transmission and distribution interconnections, with a growing number of specific applications such as large wind and solar farms along with applications relating to the use of FACTS technologies and grid battery devices. A typical family of NGCs includes connection codes, operating codes, planning codes, and market codes (Figure 11).

Figure 11: ENTSO-e Family of Grid Codes

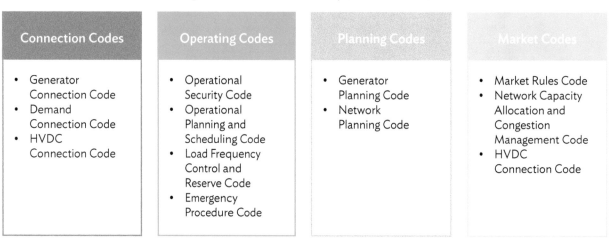

Connection Codes	Operating Codes	Planning Codes	Market Codes
• Generator Connection Code • Demand Connection Code • HVDC Connection Code	• Operational Security Code • Operational Planning and Scheduling Code • Load Frequency Control and Reserve Code • Emergency Procedure Code	• Generator Planning Code • Network Planning Code	• Market Rules Code • Network Capacity Allocation and Congestion Management Code • HVDC Connection Code

ENTSO-e = European Network of Transmission System Operators for Electricity, HVDC = high voltage direct current.

Source: International Renewable Energy Agency. 2016. *Scaling up Variable Renewable Power: The Role of Grid Codes.* May.

---

25 DNV. Power and Renewables. International Grid Code Compliance Listing. https://www.dnv.com/publications/international-grid-code-compliance-listing-138159.

RGCs are needed to increase transparency and provide equal treatment by making the same rules applicable to all. Appendix 5 lists the 12 main power trading regions around the world that have or are developing RGCs in some form or another. Typically, RGCs do not replace NGCs, but provide a common framework for grid code requirements and set minimum standards that all national grid codes must meet. Requirements for RGCs are therefore less specific and more often distinguish between countries, technologies, and synchronous zones. This approach allows the flexibility for NGCs to set tighter country-specific requirements. Like other governing frameworks, both NGCs and the RGCs need to be continuously adapted to changing technologies, system conditions, and political aspirations. RGCs are also expected to provide the overarching rules for groups of NGCs but must also be approved by national regulators before they can be enforced. For this reason, transmission system operators (TSOs) must be heavily involved in drafting and revising RGCs to avoid conflicts with NGCs. Their codes are generally published by a regional authority who then regularly consults with the other stakeholders.[26]

Adaptation of the ENTSO-e Codes in the GMS

The first European cross-border interconnections became operational in 1921 for the transmission of electricity over roughly 700 km from France via Switzerland to Italy. Cross-border cooperation on electricity has been pursued systematically in Europe since soon after World War II. In 1955, synchronous cross-border electricity exchange was possible up to a capacity of about 100 MW and electricity supply was mainly a national task. By the 1960s, a uniform 380 kV grid extended across Western and Central Europe. It provided an important instrument for effective mutual aid in the case of power system failures and seasonal trading between coal-fired versus hydro plant-dominated regions. In 2008, the Union for the Co-ordination of Production and Transmission of Electricity (UCPTE) together with other organizations merged into the European Network of Transmission System Operators for Electricity (ENTSO-e). This enabled trading at least cost on the power exchanges in the respective member countries, allowing for the optimization of generation resources through cross-border cooperation.

The underlying assumptions and trends that are driving these changes include the advent of renewables, increased electrification of industry, and decentralization of energy resources. These anticipate greater use of digitalization and secure coupling of end-use sectors as the EU shifts from being a fossil fuel-dominated and supply-centric model to a clean digitalized and electrified consumer-centric system with many distributed resources. Coordination between TSOs at the regional, synchronous area, and pan-EU levels have historically been successful developed proactively by TSOs. Since May 2016, there has been significant progress toward European grid code harmonization after the ENTSO-e became a binding EU regulation.

ENTSO-e guidelines for code development allow member countries to trade power on power exchanges in the respective countries, where any entity can trade within a span as low as 15 minutes. In 2016, the establishment of several European Regional

---

26   DNV GL (Germany) International Grid Code Compliance Listing. http://www.dnvgl.com/GridCodeListing.pdf.

Coordination Centers has enhanced the effectiveness of the ENTSO-e code and extended its scope. This umbrella European regional network code has been developed by 42 TSOs from 35 countries. While the ENTSO-e codes contain many common grid interconnection requirements, most are "non-exhaustive," meaning that the detailed specifications are still being set at a national level. In parallel, the planned creation of an EU Distribution System Operators entity (EDSO) will further strengthen coordination between TSOs and distribution system operators (DSOs) to integrate large shares of distributed energy resources.

The GMS RGC is modeled on the 50 hertz (Hz) ENTSO-e codes because there are similarities in the geographic characteristics of the region as well as historical links to international metric systems, electrotechnical standards, and system designs. Although it has been advantageous for the GMS TSOs to avail of previous experience when developing their RGCs, the ENTSO-e codes cannot be adopted word for word. For example, in contrast to the GMS power systems, the EU's tightly interconnected 380kV HVAC networks have been synchronized over many years. More recently, an increasing number of HVDC submarine cable interconnections are being connected to islanded grids in the region, and it is likely that a hybrid HVAC/HVDC EU network will emerge as power trading expands.[27]

It is unlikely that the separate GMS networks will be synchronized in the same way before grid-to-grid power trading is initiated. For system security and geopolitical reasons, GMS TSOs are wary of implementing synchronous interconnections with their neighbors while they have doubts about the security and compatibility of neighboring systems. The GMS is also pioneering developments such as retrofitted floating photovoltaic (PV) installations in the Lao PDR, the PRC, Thailand, and Viet Nam. The PRC is pioneering developments in wind generation and building HVDC super grids (800 kV and above) designed for expansion into other regional grids. Given the successful developments with multi-terminal HVDC-VSC systems in the PRC, this technology may be the best way to initiate an early form of grid-to-grid power trading. This will enable the TSOs to gain confidence with operations such as reserve sharing and VRE mitigation until such time the infrastructure and institutional arrangements are ready for regional synchronization. At that point, as recommended in the MHI Master Plan, it should be possible to pursue a more aggressive approach to developing an extensive network of HVAC interconnections.

Technical Applications of Regional Grid Codes

General Considerations

Flexible, stable, and transparent electricity power markets are considered a precondition for the transition to the modern era of electricity development,

---

[27]   ENTSO-e. ENTSO-E Transmission System Map. https://www.entsoe.eu/data/map/.

complemented by grid-to-grid power transfers. Power trading requires cross-border transmission interconnections between market platforms for wholesale, retail, balancing, and specific day-ahead and intra-day energy markets. It can also address national concerns about energy security with interconnections offering opportunities to share reserve capacity and resources in regions of economic cooperation. Distributed VRE resources can, with appropriate mitigation technologies to maintain stability and power quality, diversify the energy mix and reduce reliance on imported fossil fuels. Diversification coupled with the increased digitalization of network operations and management can also improve the flexibility and resilience of the grid to ensure the security of supply.

Power trading on a bilateral basis can be easily monitored and managed so that sudden interruptions are catered for in the provision of generation reserve capacity. Likewise, power trading using HVDC links provides TSOs with a high degree of control that can buffer problems in one system being passed through to another. On the other hand, power trading via HVAC links between synchronized grids can cause (i) potential dynamic stability issues and circulating currents in adjacent networks that result in circuit breaker tripping due to other lines being overloaded, (ii) sudden loss of synchronism that results in a prolonged shutdown and excess fault levels in one grid causing damage to a smaller grid, and (iii) metering not properly accounting for losses incurred in a participating grid.

The following subsections are not intended to cover the multitude of technical and commercial aspects that are incorporated into the RGC and elaborated in Section 5. Instead, they focus on key issues that need to be considered in the GMS region as the first grid-to-grid interconnections are made. This will include considering the impact of new technologies for interconnections (such as HVDC-VSC, VFT, and FACTS systems) and the opportunities for TSOs to procure ancillary services and energy balancing through grid-to-grid trading while addressing the concerns the TSOs have as their networks become more susceptible to adverse impacts on one grid being reflected through interconnections to adjacent grids.

Harmonization of the National Grid Code with the Regional Grid Code

Harmonization means the adjustment of differences and inconsistencies among measurements, methods, procedures, schedules, specifications, or systems to make them uniform or mutually compatible. In the case of synchronous interconnections, voltage, insulation strength, frequency and protection schemes must match. In contrast to asynchronous interconnections, a fault on one side is not easily passed on to the other, so the two sides are less concerned about the impacts on each other. Nevertheless, the RGC must cover situations where tripping an HVDC terminal or a large industry with its own in-house generation plant would constitute a disturbance in terms of loss of load or loss of supply. Accordingly, there must be real-time communication through hotlines, data transfer and cooperation between the TSOs.

Grid details must be shared between TSOs to prepare joint emergency response and recovery procedures. Mutual trust between the TSOs is therefore essential if they are to maintain and update technical data and information on the electricity sector in an agreed template on a common database. It may be emphasized that the objective of harmonization is to arrive at a practical working arrangement for secure and reliable grid operation, and it should not be construed as an attempt to impose a uniform NGC in each GMS country.

Connection Codes

The connection code is normally the largest and most technically complex single component of the family of grid codes. In most countries, it is also a mature document that has been developed over a long period. Most of the detailed technical issues deal with the interconnection of conventional power generating plants along with the management of associated levels of protection and security of the national power system. The technical characteristics of generating power plants, categorized according to their capacity to influence system operations, have evolved over the last few decades and are well-documented. Because HVDC converter installation and large industries are associated with many common harmonic disturbances that impact power quality and system stability, both these types of facilities are usually covered under the same connection code. Some aspects of the codes largely cater to new electronic technologies and are still under development. Toward this end, the RGC connection codes must be continuously updated to ensure there is a high degree of compatibility between NGCs that may have historically been derived to cater for a variety of international manufacturers.

VRE generation from large-scale wind and PV connections will have a major influence on system operations. To ensure that wind turbine generators behave more like conventional power plants with synchronous generators, TSOs in countries led by the PRC, the EU, and the United States have designed new technical standards (footnote 12). Many aspects are still being refined to cater to new developments with smart inverters and HVDC submarine cables. As discussed here, TSOs normally require VREs to contribute to power system (voltage and frequency) control and define turbine behavior, including their ability to ride through faults during grid disturbances.[28] In addition to maintaining transmission stability, special wind grid codes will also increase the transparency of technical negotiations between the power plant and transmission system operators and outline the technical parameters for wind power equipment providers. When regional power trading is in force, the overriding technical and commercial issues are required, as discussed in Section 4 to be incorporated into the framework of the RGC to ensure that network impacts of power trading are properly identified in allocating costs and benefits.

---

[28] "Fault ride through" relates to the capability of power generating modules (including DC connected power park modules) and HVDC systems to remain connected to the system and operate through periods of low voltage at the grid entry point or user system entry point caused by secured faults.

The typical problems associated with HVDC terminals and with large industries that have specialized production applications include

(i) **Voltage stabilization and flicker reduction.** These effects on the power system are caused internally by rapidly changing loads like big drives for ore mills, compressors or pumps, or electric arc furnaces and lead to undesired voltage fluctuations. Flicker in lighting can affect human well-being and degrade the performance of other connected devices.

(ii) **Harmonic distortions.** Nonlinear loads from converters or arc furnaces are a source of unwanted harmonics that could disturb consumers connected to the network. Harmonics also decrease the efficiency of transmission networks since they generate losses in all network elements, without driving a machine, melting steel or other tasks of industrial production.

(iii) **Low-power factors.** Significant inductive loads like motors are present in all fields of industry. In the metallurgical industry, arc furnaces constitute the largest inductive loads working at an extremely low power factor. The reactive power required by such loads causes degrading of the power factor. Since reactive power needs to be transported through the network, this leads to poor use of all transmission components such as cables and transformers.

(iv) **Unbalanced load.** Unsymmetrical loads are not always balanced, or at least not completely. Single-phase loads or arc furnaces can cause uneven phase currents in the three-phase supply system and result in uneven phase voltages.

Operating and Market Codes

These groups of subcodes are not usually available to power sector stakeholders in many jurisdictions partly because TSOs have developed their own sets of in-house operating rules and partly because the concept of creating a competitive market for TSO operational services is relatively new. The gap assessment (Section 5) has identified this aspect of the NGCs as one that needs considerable attention to align them with the RGC. This will ensure regional power systems can be securely synchronized to enable reliable trading arrangements between TSOs and enable stakeholders to contribute to some of the services described below.

Ancillary Services

When grid-to-grid interconnections are established, TSOs will be able to purchase ancillary services including frequency keeping, instantaneous reserve, over-frequency reserve, voltage support, and black start. The functions of the most common range of ancillary services are as follows:

(i) Frequency containment reserve (or primary reserves) must balance any generation and demand inequalities to maintain the grid frequency at or near 50 Hz under normal operating conditions and managing frequency time error. Factors that contribute to inequalities under normal operating conditions include unanticipated load changes, differences in generator ramping, and inaccuracies between the modeled and actual system conditions.

(ii)   Frequency restoration reserves (or Secondary Reserves) must manage frequency recovery after an under-frequency event, with the aim of arresting the frequency fall and recovering the frequency to 50 Hz. They help restore the Area Control Error (also known as Frequency Restoration Control Error) of each load frequency control area toward zero. In effect, they restore the system frequency to its set point (normal) value and maintain the power interchange program among load frequency control (LFC) areas.

(iii) Replacement reserves (or Tertiary Reserves) are required to manage frequency recovery after an event that might otherwise cause the grid frequency to exceed specified limits. The TSO's objective is to arrest the rise in frequency and bring it back within the normal band.

(iv)   Voltage support is required to provide additional reactive power resources of the static or dynamic type, depending on the location and network loading conditions, to contribute to network voltage control when dispatched. This is normally a local technical issue, which is typically mandatory and not remunerated. The volt/VAR control process is implemented by manual or automatic control actions designed to maintain the nominal set values for the voltage levels and reactive powers. Usually, there are also bilateral contracts held between a service provider and a TSO for the provision of extra volt or VAR control.

(v)   Black start capability is required to maintain equipment that can initialize the supply for the progressive relieving of the grid following a partial or total blackout. The service is intended to power up other plants and loads to bring the grid system back to normal operating conditions. Black start is normally contracted through bilateral agreements and is not expected to have its own market framework.

The development of a wholesale market in Viet Nam provides an opportunity to consider a controlled HVDC interconnection with Thailand to create a common mode for ancillary services initially with each other and eventually within the region. In future and as more VRE comes on line, GMS TSOs can also be expected to establish and procure new applications such as fast frequency response, synchronous inertia, synthetic inertia, short- and long-term storage, and demand side management support.

Reliability Standards

Reliability standards are determined using economic considerations, including the cost to consumers of losses of supply while meeting technical and safety requirements. The RGC planning code contains principles for their development and what must be included in the standards. They include an economic (probabilistic) standard for the regional grid and the associated assessment of the costs and benefits of investment for reliability and a safety net minimum reliability standard of N-1 for contingencies on the core parts of the national grids.[29] For parts of the system requiring even greater security, an N-2 standard is sometimes required.

---

[29]   The N-1 criterion is a generally accepted rule according to which elements remaining in operation after the failure of a single network element (such as a transmission line, transformer or generating unit, or in certain instances a busbar) must be capable of accommodating the change of flows in the network caused by that single failure.

However, for grid-to-grid interconnections, the reliability criteria is not an equivalent measure for every transmission circuit's reliability nor is it applicable for all interconnection configurations with different structures and lengths traversing types of terrain and where the lines are exposed to adverse environmental conditions. For example, a double circuit HVAC transmission line has large heavy towers that are not quickly replaceable if damaged by adverse weather or terrorism. The towers carry six conductors, where any one conductor or insulator damaged by adverse weather can result in disruption of a (three conductor) circuit. On the other hand, a double circuit HVDC line with converters at its terminals is less affected by adverse weather events. This is because it has lighter, more easily replaceable structures and has only two conductors exposed to the elements along with an extremely high reliability backup ground return system. To achieve comparable N-1 service with HVDC lines, for configurations using a back-to-back converter connected to a long double circuit HVAC line, it may be necessary to build a third (single) circuit HVAC line in parallel with the main line.

Interconnection terminations are normally selected at substations where two or more lines are already in place to ensure both a reliable point of supply and evacuation. Accordingly, for a 2,000 MW HVAC substation comprising four 500 MW transformers, there may well be a case for providing N-1 capability within the substation. This is achieved by installing a spare 500-megavolt-ampere (MVA) transformer (given that outdoor transformers are prone to environmental impacts like overheating, lightning, oil leaks, fires, etc.). In comparison, two HVAC/HVDC converters are up to five times the cost of a conventional substation but are generally more reliable than transformers since they are housed in their own buildings, which are less prone to environmental problems.

Contingency Planning

The ultimate objective of an interconnection, like the power systems it is part of, is to provide power to customers economically, safely, reliably, efficiently, and with minimal negative environmental impact. Each of these aspects has one or more quantitative measures such as price/kWh, number and lethality of accidents, frequency, and duration of service interruptions, plant heat rate, transmission and distribution losses, and emissions factors. Interconnections are designed and their components selected with these objectives in mind, though they would be optimized differently in different systems. TSOs of interconnected systems are obliged to monitor the N-1 principle not only for their own grid but also for the tie-lines to neighboring grids.

TSOs must use system planning studies based on deterministic analysis, in which several representative planning cases are considered. This approach aims to assess the likelihood of risks of grid operation throughout the year and to identify the uncertainties that characterize it. The aim is to cover many transmission system states throughout the year by creating multiple cases depending on the variation of uncertain variables. Planning scenarios are also performed to represent future developments of the energy system. The essence of scenario analysis is to come up with a plausible picture of the future. Scenarios are means to approach the uncertainties and the interaction between these uncertainties. Planning scenarios are coherent, comprehensive, and internally consistent descriptions of a plausible future characterized by several time horizons, technical parameters, economic parameters, generation portfolios, demand forecasts, and exchange patterns.

Due to the increase of interconnections among TSOs, the assessment of security is more and more interdependent and dependent on load and transmission contingencies as shown in Figure 12.[30]

This characteristic of power system operations mandates the TSO to consider the influence of the surrounding grid on its "responsibility area." For this purpose, an external contingency list is prepared that includes all the elements of surrounding areas that influence its responsibility area which are higher than a value called the "contingency influence threshold." That means the branches of the external contingency list must be modeled along with surrounding branches with lower influence on the responsibility area, and must be part of the model that changes as shown below. This will ensure accurate simulations of the effects of external outages. All the external elements with an influence on the responsibility area higher than a certain value called the "observability influence threshold" constitute the external observability list. Consequently, all TSOs establish an individual grid model on a year-ahead basis and update it on a week-ahead basis or more often as needed. Apart from the system topology modeled down to 110 kV, the individual grid model includes the thermal limits, voltage limits, stability limits, and fault current thresholds.

Figure 12: Impact of Power Flows on the Shape
of the TSO Voltage Control Area

Voltage Control Area - Changing Shape and Size with Load Levels and Transmission Contingencies

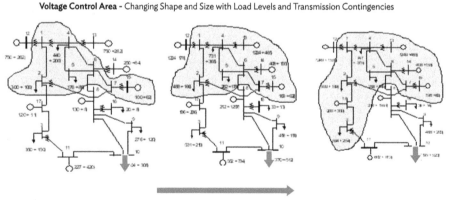

Load Increase and Transmission Contingencies

TSO = transmission system operator.
Source: Electric Power Research Institute (EPRI). 2007. *EPRI 2006 Annual Report.*

---

30   Electric Power Research Institute (EPRI). 2007. EPRI 2006 Annual Report.

Metering

Power trading in the GMS will require a correct assessment of both the wheeling transactions and the impacts of active and reactive power flows on the intermediate grids. As the number of transactions increases, the grids will require added metering, telemetry, and communication for load frequency control systems. At a later stage of development, a secondary model could emerge, involving the development of a GMS regional market existing separately from national markets and system operations. This would build on some elements set up in the harmonized bilateral model, such as harmonized wheeling charges, and introduce new elements, such as a regional market operator and a central clearing mechanism. Eventually, GMS countries may even choose to replace their national markets with a fully integrated regional market to bring added benefits to participating countries, but this would also require more changes at the national level, including market restructuring.

For this first phase of grid-to-grid power trading, metering of wheeling will be needed between control areas as well as within them. As the volume of power trading transactions increases, the control centers of the transporting utility will also need upgrading and more dispatchers may be needed. Metering tie-lines into each area is an integral part of these systems and capable of measuring flow in both directions, with energy and capacity measured consistent with the requirements of the affected systems. Implementing large numbers of wheeling transactions may require TSOs to revisit the concept of control areas and load frequency control. This is because large numbers of wheeling transactions would strain the TSOs to the extent that wheeled power is used to continuously balance load and supply for many retail customers or small full-requirement utilities. As loads and generators become independent from integrated utilities under retail wheeling or wheeling to requirements utilities, they can, in concept, become separate control areas buying and selling power using interchanges with the transmission system.

Communications between National Centers

The principal responsibilities of the TSO relate to system security, planning and scheduling of regional power trading, load frequency control and reserve management, emergency and restoration planning, and market and metering issues. This entails continuously monitoring system conditions and deploying system resources as the situation requires. Increasingly, these activities are automated. Supervisory control and data acquisition systems (SCADA) combine remote sensing of system conditions with remote control over operations. SCADA systems control key generators through automatic generator control and can change the topology of the transmission and distribution network by remotely opening or closing circuit breakers enabled by dedicated communication systems (often fiber optic), microwave radio, and power line carrier signals.

The importance of communication protocols such as the ENTSO-e Common Information Model for energy data exchange in the wider decarbonization agenda cannot be underestimated. As the GMS utilities move from centralized fossil to distributed renewable generation coupled with extending smart grid deployment, a significant increase of process data needs to be exchanged in real-time between systems and applications of many different participants to actively manage the complexity of the network and energy balance. Digitalization in the power industry is not new. The industry has a history of applying digital tools and upgrading old analog systems. This includes the automation of network operations and the emergence of sensors and data analytics to enable smart asset management leading to better use of main system elements. Seamless and efficient exchange of information is necessary at stages between an increasing number of companies, TSOs, DSOs, and generators.[31] Such information exchanges have become indispensable in network planning, power system operation, and market facilitation.

Emerging Issues

Cybersecurity

With the rapid development of the smart grid and increasingly integrated communication networks, power grids are facing serious cybersecurity problems. A recent technical paper has reviewed studies on the impact of false data injection attacks on power systems from three aspects.[32] First, false data injection can adversely affect economic dispatch by increasing the operational cost of the power system or causing sequential overloads and even outages. Second, attackers can inject false data into the power system state estimator, and this will prevent the operators from obtaining the true operating conditions of the system. Third, false data injection attacks can degrade the distributed control of distributed generators or microgrids inducing a power imbalance between supply and demand. Both ENTSO-e and NREL are addressing the problem and are developing a network code to deal with these concerns.[33]

Confidentiality of Data Exchange

Inter-TSO cooperation will significantly affect data and information exchange. There are both semi-static (such as system topology) and dynamic (such as real-time operating data) data and information to be exchanged much of which is commercially sensitive and confidential. This has proved to be a problem area that needs to be resolved to enable power trading to function effectively. Although there are sections in the RGC concerning confidentiality obligations in some subcodes (such as in Load Frequency Control and Reserve Code, Operational Security Code, and Operational Planning and Scheduling Code), they are considered by some TSOs as too general and somewhat weak. Hence, there appears to be a need for a separate Confidentiality

---

[31] *Network Magazine.* 2016. What Does DSO Really Mean? 29 November. https://networks.online/heat/what-does-dso-really-mean/.

[32] Y. Xu. 2020. A Review of Cyber Security Risks of Power Systems: From Static to Dynamic False Data Attacks. *Protection and Control of Modern Power Systems.* 5 (19). 4 September. https://doi.org/10.1186/s41601-020-00164-w.

[33] ENTSO-E. 2021. Recommendations for the European Commission on a Network Code on Cybersecurity.

Obligation Code that identifies items of confidential data not to be disclosed without the owner's consent, specified procedures for how to obtain permission to share data or information with third parties (if necessary), and the penalties for confidentiality violations.

Optimization of Energy Storage

There are emerging battery and mechanical electricity storage devices that can operate in a similar way to a hybrid combination of a generator and load.[34] The value of such energy storage systems in providing fast frequency response has been recognized for their ability to support peaking and demand-side management services. In some cases, battery electric storage systems (BESS) already participate in wholesale electricity markets as either generation and dispatchable demand but generally cannot offer instantaneous reserve because the existing national codes are technology specific. In recent years, the capital cost of BESS has decreased drastically, while their technology, reliability, and life cycle have increased significantly.[35]

As the volume of renewables grows, energy storage will become important for managing supply and demand. Apart from providing ancillary services, the impact of storage in wholesale energy markets depends on how much and how quickly fossil-based power generation is phased out. Adding new storage technologies to the grid may result in different fault conditions, moving from classical short circuit currents to short high-current pulses. Using flexible hydropower and pumped storage is on the agenda for large-scale storage but the commercial viability of this option is very country dependent. In this respect, large hydropower occupies a unique position because it can store primary energy (GWh) with high efficiency as the potential energy of water. Importantly, it can also provide power capacities gigawatt [GW] at a high degree of predictable availability.

The rapid growth of electric vehicles (EVs) is also expected to help both TSOs and DSOs cope with this increasing need for flexibility in the system. High numbers of EVs represent significant battery storage capacity and are therefore the perfect companion to renewables. Energy suppliers could save money through portfolio optimization, for example, by optimizing the charging and discharging cycles of the EV against intra-day market price movements. The real value will be in developing new business models that will capture these potential benefits. There are already energy suppliers offering products to consumers based on vehicle-to-grid schemes, such as free charging for owners of EVs. For the energy supplier, it is also attractive because it is a way for them to build long-term relationships with a group of high-end consumers.

Although the development of a variety of new storage technologies has made them technically feasible, to be integrated on a larger scale with required performance,

[34] L. Meng et al. 2020. Fast Frequency Response From Energy Storage Systems–A Review of Grid Standards, Projects and Technical Issues. *IEEE Transactions on Smart Grid.* 11 (2). pp. 1566--1581. https://doi.org/10.1109/TSG.2019.2940173.

[35] E. Rakhshani et al. 2019. Integration of Large Scale PV-Based Generation Into Power Systems: A Survey. *Energies* 12 (8) 1425. https://doi.org/10.3390/en12081425.

the policies, grid codes, and procurement issues are still presenting barriers for wider application and investment. In the GMS context, this means that new regulatory proceedings at the national and state levels may be needed to enable energy storage projects to participate as a source of both load and generation and to provide multiple grid services. For utility-owned energy storage devices, where costs are recovered under a cost-of-service regulation, utilities and regulators can establish agreed-upon methods to quantify and compensate the full system value that energy storage provides to the power system.[36]

Variable Renewable Energy Expansion

Plans for expanding the use of renewable energy in the GMS region raise several technical and economic issues that must be considered in the RGC. In this regard, the International Renewable Energy Agency (IRENA) published in 2016 comprehensive guidelines for the development of grid codes based on experience in countries planning to increase the uptake of VRE capacity.[37] The report also sets out a range of issues to be considered in the design and implementation of grid codes including voltage and frequency operation ranges, power quality, reactive power capability for voltage control, frequency support, fault behaviors, active power management and gradient limitations, simulation modeling, communications, protection as well as new applications relating to the contribution of synthetic inertia, black start capability, and damping of power system oscillations. VRE schemes for large-scale PV solar and offshore wind farms with HVDC cables require grid interconnections via large inverters where HVDC connection codes would apply. In cases such as floating PV installations, there is a combination of HVDC and HVAC production that may incorporate large hydraulic or battery storage systems to match generation supply to demand curves. A publicly available comprehensive technical guide for VRE generation facilities has recently been published that covers grid integration requirements, compensation devices, studies, and forecasting systems.[38] This document also discusses the technology in general and makes recommendations for VRE technical specifications, applicable standards, and essential testing. It will help practitioners understand some of the essential requirements and available technical and regulatory measures to integrate large shares of VRE into power grids without compromising the adequacy, reliability, or affordability of electricity.[39] In 2019, IRENA issued a second report with insights to how VRE technologies might affect the development of future grid codes.[40] This includes chapters on advanced

[36]   I. Chernyakhovskiy et al. 2021. *Energy Storage in South Asia: Understanding the Role of Grid-Connected Energy Storage in South Asia's Power Sector Transformation*. NREL. https://www.nrel.gov/docs/fy21osti/79915.pdf.

[37]   IRENA. 2016. *Scaling up Variable Renewable Power: The Role of Grid Codes*. May. https://www.irena.org/-/media/Files/IRENA/Agency/Publication/2016/IRENA_Grid_Codes_2016.pdf.

[38]   Energy Sector Management Assistance Program. 2019. *Grid Integration Requirements for Variable Renewable Energy: ESMAP Technical Guide*. Washington, DC: World Bank. https://openknowledge.worldbank.org/handle/10986/32075.

[39]   D. Brown. 2019. Analysis: Floating Solar Power Along the Dammed-Up Mekong River. *Mongabay*. 3 December. https://news.mongabay.com/2019/12/analysis-floating-solar-power-along-the-dammed-up-mekong-delta/.

[40]   EPRI. 2019. *Meeting the Challenges of Declining System Inertia*. https://www.epri.com/research/products/000000003002015131

weather forecasting, flexible generation to accommodate variability, interconnections, regional markets as flexibility providers matching renewable energy generation and demand over large distances with super grids, large-scale storage, and new grid operation to defer grid reinforcements investments.

Inertia Management

Inertia in power systems refers to the energy stored in large rotating generators and some industrial motors, which gives them the tendency to remain rotating. This stored energy can be particularly valuable when a large power plant fails, as it can temporarily make up for the power lost from the failed generator. This temporary response, which is typically available for a few seconds, allows the mechanical systems that control most power plants time to detect and respond to the failure. The replacement of conventional generators with inverter-based resources, including wind, solar, and certain types of energy storage have two counterbalancing effects. First, these resources decrease the amount of inertia available. Second, they can respond much faster than conventional power plants, reducing the amount of inertia needed and thus addressing the first impact. In combination, this is a change in thinking in how TSOs will consider alternatives to providing frequency response. The combination of inertia and mechanical frequency response is expected to be replaced to a large extent with electronic-based frequency response from inverter-based resources and fast response from loads while maintaining system reliability.

A recent EPRI white paper reviews the nature of synchronous inertia, addresses the factors leading to declining inertia, and examines options for ensuring stable system operation (footnote 41). The authors surveyed the technical and economic issues that arise from operation under reduced inertia. The paper concludes that the power industry needs new analytical tools as well as high-quality real-world data on the effects of reduced system inertia during disturbances. In the meantime, new techniques for supporting system inertia require study to determine their value and effectiveness in supplementing or replacing synchronous inertia. These techniques include markets for frequency services and technological approaches to emulate the effects of inertia.

5. Scope of the GMS Regional Grid Codes

Overview of the GMS Regional Grid Codes

Current Status

The GMS RGC is available on the ADB GMS website.[41] As summarized in Table 7, it comprises a document of 436 pages in 11 parts, including a 44-page glossary of terms. In keeping with convention, the RGC is structured according to the ENTSO-e family of codes (Figure 11) relating to regulations for connection, operations and markets.[42] In contrast to the GMS RGC, the ENTSO-e codes are stand-alone legal documents, each with its own preamble (beginning "WHEREAS" and finishing "HAS ADOPTED THIS CODE" with each subsection incorporating a penultimate "Article on Operational Training and Certification").

Table 5 lists the seven specific codes (and subcodes) that make up the total set of GMS codes by volume number, code title, and other particulars.

The documents also include a GMS Strategic Planning Code to establish the common limits of equipment capacity. As transmission systems of the GMS countries become synchronously connected, it may be necessary to limit the impact of the higher fault levels of smaller countries to avoid having to change their circuit breakers because of the relative differences in national power system capacities. This can be achieved by having fault current limiters at the interconnecting substations of the neighboring countries.

The GMS RGC applies to both HVAC synchronous and HVDC asynchronous interconnections, although the requirements for compliance with synchronous connections is a key objective. In both cases, common scheduling of time intervals and restoration plans are needed in case of a grid failure. In synchronous connections, there is an additional requirement for a common frequency range along with many other stipulations. However, the scope of the GMS RGC is much less than the full-fledged NGCs that go into more detail as required for technical design, compliance, and registration. As noted in Section 4, the NGC's can be expected to contain many common grid interconnection requirements since the detailed specifications should still be set at a national level.

---

[41] GMS. 2021. Greater Mekong Subregion Regional Grid Code. https://greatermekong.org/greater-mekong-subregion-regional-grid-code.

[42] ENTSO-e. What Are Network Codes? https://www.entsoe.eu/network_codes/.

Table 5: Summary of GMS Regional Grid Codes

Volume	Code Title	Summary of Code Particulars
1	Preamble V0.4 December 2018	Context and objectives of RGC detailed summaries of each of the succeeding sections
2	Governance V0.4 December 2018	Provisions necessary for the overall administration and review of aspects of the RGC
3	Connection V0.4 December 2018	Connection conditions for power-generator facilities, HVDC systems, including DC connected power modules and demand facilities
4	Operational Security V 0.2 December 2018	Principles for transmission systems applicable to all TSOs, DSOs, and significant grid users in normal and alert system state
5	Operational Planning and Scheduling V0.2 December 2018	Requirements for ensuring coherent and coordinated operational planning processes of Synchronous Areas
6	Load Frequency Control and Reserves V0.1 December 2018	Minimal requirements and principles for load frequency control and reserves
7	Emergency & Restoration V0.2 December 2019	Operational security requirements and principles applicable for emergency state, blackout state, and restoration
8	Market V.02 December 2018	Capacity allocation and congestion management, forward capacity allocation, and electricity balancing
9	Metering V0.2 December 2018	Minimum technical, design, and operational criteria to be complied with for the metering of each point of interchange of energy between control areas
10	Operational Training V0.2 September 2018	Framework for operational training for dispatchers to always operate the power system in a safe and reliable manner under all conditions
11	Glossary of Terms	List of terms, acronyms, and units commonly used in the GMS Transmission Regulations Policy 1 to 4, and the GMS Grid Code
12	GMS Strategic Planning Document	Guidelines for medium- and long-term planning and specification of planning data required

DC = direct current, DSO = distribution service operator, GMS = Greater Mekong Subregion, HVDC = high voltage direct current, RGC = regional grid code, TSO = transmission system operator.

Source: GMS. 2021. Greater Mekong Subregion Regional Grid Code.

The GMS RGC envisages the installation of data acquisition systems, disturbance recorders, and sequence-of-events recorders at the interconnection of significant points to enable protection coordination between grids. This will ensure robust, redundant, and reliable communication between grids so that voice and data communication can take place instantly and seamlessly across countries. These provisions could be set out in a separate Information Exchange Code as indicated in the Governance Code Section 2(g). Once the RGC is adopted by all TSOs and enforced by their regulators, power trading should be able to start increasing with the respective national grids operating securely and efficiently.

Summary of RPTCC Comments

The GMS RGC documents include annexed comments and observations from the working group members, along with responses from moderators that were recorded at the relevant RPTCC meetings where they were discussed and approved. This practice is in accordance with the provisions of Governance Code 7.3(1), which requires the RPCC to have a "clear expression of divergent views on such proposals if any were received." A summary of the comments and resolutions is listed in Table 6.

Table 6: Summary of RPTCC Member Comments on RGC and Recommended Resolutions

RGC Section	Page	Month Year	PRC	THA	VIE	CAM	MYA	LAO	Remarks
Preamble	11	12/18	1		5			3	Minor observations
Governance	23	4/18	1	2	2			1	Text changes made
Connections	94	6/18	10	7	25				Technical differences
Operational Security	49	8/18	10	3	6				TSOs to resolve
Operational Planning and Scheduling	47	12/18	7	7	2				TSOs to resolve
LFC and Reserves	64	4/18							No comments
Emergency & Restoration	42	8/18	5	1					TSOs to resolve
Market Code	20	4/18		4					Minor modifications
Metering	13	8/18	1	1	1				RPTCC 24 Comments
Operational Training	12	9/18	3						Minor additions
Glossary	50	12/18							No comments
Strategic Planning	11	12/18							No comments

CAM = Cambodia, GMS = Greater Mekong Subregion, LAO = Lao People's Democratic Republic, LFC = load frequency control, MYA = Myanmar, PRC = People's Republic of China, RGC = regional grid code, RPTCC = Regional Power Trade Coordination Committee, THA = Thailand, TSO = transmission system operator, VIE = Viet Nam.

Source: GMS. 2021. Greater Mekong Subregion Regional Grid Code.

Gap Assessment and RGC Implementation Strategy

In November 2019, a completion report relating to the designated four tasks assigned to the WGPO was prepared to describe the WGPO achievements and its plans for implementation. The report summarizes the status of a regional power trade framework in terms of the policy and institutional measures in place and the intergovernment agreements including the three MOUs issued in 2005, 2008, and 2012. It describes the status of the organizational and institutional regulatory frameworks for each GMS member in terms of each country's energy institutions, regulatory authorities, TSOs, and their associated NGCs. It concludes that the RPTCC has no legacy or mandate to enforce the RGC at the regional level any decision regarding the establishment of a regional power trade organization. The RPTCC can only report to the GMS Ministerial Level Conference and the respective national governments through the responsible minister who has the power to make the RGC enforceable at the national level after consideration of the obligations given to the national regulatory authorities. To achieve this, it appears another MOU may be required to establish a GMS Grid Code Secretariat and Review Panel.

The report considers the national and regional prerequisites relating to institutional, regulatory, and technical considerations to facilitate open access for transborder power exchange. At a national level, it recommends there must be a strong and sustained commitment to liberalization and competition in the energy sector defining the GMS governments intention toward third-party access. Accordingly, the report concludes that new legislation would be needed to mandate the general principles of open access, competition, and non-discriminatory access to the national grid. This would need to be supplemented with regulations covering licensing, tariff methodology, customers, accounts, and complaints along with changes to the distribution codes. Legislation is also required to ensure a strong and credible regulatory body, an independent TSO, and a transparent planning system with the free exchange of information subject to a code of confidentiality between TSOs and the RPTCC.

At a regional level, there are also prerequisites that need to be addressed by the RPTCC relating to the requirements for the development of regulation codes and guidelines to enable a regional market structure. These include the development of eligibility criteria and guidelines for registration of market participants, tariff methodologies, including a model PPA for all new entrants. In addition, there are technical, financial, and tariff prerequisites designed to safeguard the integrity of cross-border transactions; regulate compliance with the RGC; and ensure there is a transparent, fair, and cost-effective tariff methodology in place.

The report includes a gap assessment of NGCs in the GMS to determine what items need to be amended to align with the overarching provision of the GMS RGC.[43] Most

---

[43] The November 2019 Gap Assessment report does not include an assessment of PRC NGC, which is thought to be compliant with ENTSO-e. The subsequent June 2021 Progress Report provides explicit information on gaps in the NGCs for Cambodia and the Lao PDR and includes comprehensive explanations of the reasons for achieving compliance with the RGC.

of the gaps identified are related to a few technical inconsistencies and the lack of provision in the NGCs mainly about operational security and market issues. In many cases, provisions are identified as "need to be added," or "necessary updates" although there are no specificity or cost implications required to ensure compatibility with the RGC. They are largely related to the codes that govern operational planning and scheduling, load frequency control and reserves, emergency and restoration, capacity allocation and congestion management, and electricity balancing. The gaps primarily relate to incompatible frequency control standards. Ambiguities and omissions in the NGCs are explained in more detail in the relevant code sections described here. In 2020, Viet Nam carried out its own comprehensive review of its grid codes by comparing them clause by clause with the ENTSO-e subcodes (Section 2). Its report recommends that the existing clauses for the many subcodes be regrouped in a similar format to the RGC to form a single new NGC document.

The report includes recommendations regarding the enforcement of the RGC in terms of the legal and regulatory prerequisites along with the technical and contractual prerequisites for enabling open access to cross-border power exchange. These are defined in Table 7 as a set of six further tasks that need to be addressed to move to the next stage of development.

Table 7: Summary of Proposed WGPO Work Plan

Group and Task Numbers		WGPO Activity	Comments
Preparation of RGC	1–4	Performance standards, transmission regulations, metering and RGC	Achieved
Compliance and Operational practices	5	Roadmap for the implementation of the GMS Grid Code	Ongoing
	6	Organization for the operationalization of the GMS synchronous areas	Review of Lao PDR, Cambodia NGCs
Regional IT system and architecture of the regional metering system	7	Design of the regional ITC system for power exchange	TA required for system design
	8	Organization and architecture of the regional metering system	Ongoing
Regional Planning Activities	9	GMS strategic planning organization for updating GMS Master Plan	Review of MHI Master Plan
	10	Assessment and management of regional transmission projects portfolio	

GMS = Greater Mekong Subregion, IT = information technology, ITC = Inter-TSO comp, LAO = Lao People's Democratic Republic, MHI = Manitoba Hydro International, NGC = national grid code, RGC = regional grid code, TA = technical assistance.
Source: ADB. 2020. Summary of Recommended Actions Described in the RGC Consultant's Progress Report V8: Section 4 WGPO Tasks 5–10.

GMS Planning Parameters

The GMS performance standards common to all the GMS codes are summarized for grid-to-grid interconnections in Table 8.

Table 8: Summary of Principal GMS Performance Standards

Voltage Control		Operating Conditions
Requirements	GMS RGC	
Normal condition		
Normal condition + Nominal V <230 kV + Nominal V >230 kV	±10% +5%,-!0%	• Voltage control is an ancillary service; all participants are obliged to provide minimum requirements as established in the grid code. • The TSO shall endeavor to maintain sufficient availability of dynamic and static reactive power for maintaining transmission system voltages at connection points within the levels specified. • MW flow limitation across interface to protect system from large voltage drops caused by contingency.
Emergency Situation	±10%	
Reactive Transfer Limit Calculation time	5% 5 minutes	

Power Plant Frequency Range		
Nominal Frequency	50 Hz ± 50 MHz	
Capability Active Power Active/Reactive Power Generator Terminals + Lagging + Leading	49.5–50.5 Hz 49.5–50.5 Hz 0.85 0.95	Continuous rating within normal voltage variations. Able to supply active or reactive power within leading and lagging limits.
Frequency Limits 51.5-52.0 Hz 51.0-51.5 Hz 49.0-51.0 Hz 47.5-49.0 Hz 47.0-47.5 Hz	Time 15 minutes 30 minutes Continuous 30 minutes 15 minutes	Design of user's plant must enable operation of generating units within the following ranges. Frequency range for synchronization.

Primary Frequency Control – Power Plants		
Af Activation	±20 mHz	Primary control
Generation Capacity	>40 MW	Must participate with governor free mode
Steady State Deviation	±200 mHz	Maximum limit of frequency deviation
Dynamic Frequency Deviation	±800 mHz	Maximum limit of dynamic frequency deviation
Minimum Maximum	<47,5 Hz >51.5 Hz	Automatic disconnections without time delay and safeguard auxiliary service supplies

Continued on the next page

Table 8 continued

Primary Frequency Control – National Control Center		Operating Conditions
Deployment time	15 seconds	Primary Control Reserves must be continuously available without interruption and regularly checked.
50% of total PCR	30 seconds	
50–100%	3%	Minimum duration: 15 minutes
Limit of loss of PCR	2.5% of peak	Percent of reference incidence size on unit tripping.
GMS Primary reserve	capacity	Each system to contribute to the correction of a disturbance pro rata to its share of annual GMS energy generation.

Secondary Frequency Control – Power Plants		
All power plants to contribute. Data to be made available to TSO related to function of contributing units.	G>100 MW	Generators, when synchronized, must reserve a proportion of power out or secondary reserve. By agreement with TSO, generators can restrict governor action where essential for safety and to avoid damage of plant. No time delays or frequency dead bands to be applied to governor control systems.

Secondary Frequency Control – National Control Center		
System Frequency Margin at set point within	A=±20 mHz 900 seconds (15 minutes)	NGC to be equipped with centralized secondary controller to change AGC or ACE set points to restore frequency and primary control reserve NGC to establish and monitor dispatch and scheduling of generating units, interchanges and outage coordination

Tertiary Frequency Control – National Control Center		
Minimum Participation	10 MW	Capacity available by all grid-connected producers and consumers to participate in balancing market.
Available capacity for activation	= Loss of largest unit in 15 minutes	Minimum capacity TSOs must always be available within 15 minutes as secondary and tertiary reserve. TSOs to establish a balancing market in accordance with RGC

Faults and Harmonic Distortion Limits				
Transmission System Voltages	Fault Clearance Times	Short Circuit Current Levels	Harmonic Voltage Distortions (imbalance 1%)	Harmonic Current Distortions (imbalance 1%)
• 500 kV	80 ms	50 kA	1–1.5%	1–1.5%
• 220–230 kV	100–120 ms	40 kA	2–2.5%	2–2.5%
• 115–132 kV	100–150 ms	31.5 KA	2–2.5%	2–2.5%

ACE = area control error, AGC = automatic generation control, GMS = Greater Mekong Subregion, Hz = hertz, kA = kiloampere, kV = kilovolt, mHz = megahertz, ms = millisecond, MW = megawatt, NGC = national grid code, RGC = regional grid code, TSO = transmission system operator.

Source: Adapted from M. Caubet. 2017. RPTCC 22 Summary of Performance Standards and Transmission Regulation: Way Forward and Work Plan. 20 June.

Connection Ratings

The NGCs specify the technical and operational requirements for the interconnection of power generation plants, DSOs and industrial demand centers for the different parties involved in the production, transportation, and use of electric power. They define the response characteristics of conventional large thermal, nuclear, or hydro generation plants, particularly during sudden changes due to loss of active power or loss of demand due to faults in the grid. Table 9 shows the GMS RGC standardized definitions of connection types and associated functionalities depending on transmission grid region, connection level, generator capacity, and technology for RGC and NGCs.[44]

The RGCs address only the most significant technical concerns associated with the transmission grid, including HVDC interconnections where compliance is required to guarantee safe operation and performance. They are typically defined to apply at the point where a facility like an HVDC terminal or a wind park is connected to the grid. Such interconnections can be part of a scheme involving a large VRE source and a large hydro source with a significant reservoir capacity or with pump storage capability. This configuration can be useful for sharing reserves between grids or for providing load shifting capability to optimize the investment in VRE sources. As wind or solar farms continue to increase in capacity, the impact of VRE generation intermittency can cause problems for a grid, particularly when there is a sudden change of capacity or loading on either side of the interconnection. In this respect, TSOs need to have overriding control of ride-through capability to ensure that large farms can wait out a disturbance and return online when the situation is stabilized.

Table 9: Standardized Definitions of Grid Connection Types

Type	Max MW	Voltage	Functionality
A	<1	LV/MV	Basic capability to withstand widescale critical events Limited automated response and control
B	1-40	<110 kV	Automated dynamic response and resilience System operator control
C	40-75	<110 kV	Stable and controllable dynamic response Covering all operational network states
D	>75	>110 kV	Widescale network operation and stability Balancing services

kV = kilovolt, LV = low voltage, MV = medium voltage, MW = megawatt.
Source: GMS. 2021. Greater Mekong Subregion Regional Grid Code. Connection Code. Para 2.1 (2).

---

[44]   The same categories apply to interconnections with generators, HVDC sources, and large industrial consumers.

Regional Grid Code Preamble and Glossary

A summary of the overall scope of the GMS RGC is given in its preamble. This explains the context of the GMS RGC in terms of plans for regional power trading in the GMS, the policy objectives, the relationship between the RGC and the NGCs, and the provision for separate synchronous zones in the transition period. The preamble establishes that the GMS RGC will take precedence over NGCs in all matters related to interconnected networks. It acknowledges that the RGC derives its legal authority from the intergovernmental agreement and provides an overview of the main sections that include the codes for governance, connection, operation, metering and system operator training. For some reason, the preamble makes no mention of the market code or the planning code that also form important parts of the RGC. The accompanying glossary code is a substantive document of 48 pages that lists alphabetically 350 internationally accepted terms, acronyms, and units used in the transmission regulations and the GMS RGC. This aggregates all the definitions included within each subcode to provide one comprehensive list for easy reference.

The nine comments recorded by the RPTCC members mostly relate to issues of clarification of the RGC being adapted to reflect the respective country strategies relating to long-term planning. It would be helpful to include more detail in the introductory section of each of the following codes that relate to the legal standing of the document, the target audience, the responsibilities of the TSOs, the role of regulators, the origin of the technical specifications, and the specific purpose of its implementation along with other supporting information.

The Governance Code

The 23-page Governance Code (GC) expands on the Articles of the MOU (2012) clarifying the objectives and functions of the RPCC governance structure comprising a board, an executive director, technical groups, and the administration. The GC provides explicit details of the responsibilities of each functional group and how they should report through the executive director to the RPCC board. This will help establish the RPCC and provide certainty to the working groups in making further progress with the implementation of the GMS RGC to expand power trading.

The GC describes the processes to be followed by a Grid Code Secretariat in updating the RGC to improve safety, reliability, and operational standards. It sets out how parties can influence the amendment process and defines who has the authority to recommend and ultimately approve and enforce changes. The GC also elaborates on the oversight and compliance provisions that need to be observed. It sets out dispute management procedures, explains how outcomes should be determined by the RPCC board and how violations and sanctions should be administered by the RPTCC. Most of the six comments recorded by the RPTCC members relate to issues of clarification.

The stated scope of the GC (para 2.3) needs to be updated to include the market code and to clarify the status of the planning code. The GC could also be enhanced by including formal articles of association for the RPCC and internal regulations based on the ENTSO-e governance structure.[45] This would govern the operation of the RPCC, its membership, the roles and relationships between the bodies and the distribution of voting rights between the members.

The Connection Code

At 94 pages, the GMS Connection Code (CC) is the largest and by far the most detailed technical section of the GMS RGC. It separately sets out requirements for generators, HVDC, and demand connections. The objective of the CC is to ensure that by specifying minimum criteria, the basic rules for connection of interconnected systems are the same for all participants of an equivalent category. GMS TSOs with tighter parameters stipulated in their national codes and that fall within the bands proposed in the RGC are considered compliant. The GMS CC will enable the maintenance, preservation, and restoration of system security to facilitate proper functioning and achieve cost efficiencies in the internal electricity market within and between synchronous areas.

Each of the three specified connection types includes separate sections defining technical requirements in terms of (i) frequency tolerance, active power, and frequency control requirements; (ii) voltage tolerance, voltage control and reactive power provision; (iii) fault ride-through capability; (iv) protection requirements; (v) system restoration, islanding and black start capability; (vi) information requirements; and (vii) connection and testing. The GMS CC includes a citation that details technical specifications for the code and closely follows EU Commission Regulations 2016/631 (Requirements for Generators), 2016/1447 (HVDC), and 2016/1388 (Demand).[46] Unlike the EU regulations, the GMS CC does not include a preamble section explaining the rationale for the selection of criteria relating to the capability of automatic response and voltage level for Types A, B, C, and D generator units.

There were 42 comments from RPTCC members representing the PRC, Thailand, and Viet Nam. The seven comments from Thailand were concerned with the clarification of under-frequency tripping issues where they considered the document was more detailed than expected. No changes were made to the specified over-frequency limit of 51 Hz since these are in line with IEC standards for the manufacture of generator plants. The 10 comments from the PRC resulted in several changes in the CC relating to the need for tighter protection requirements, the need for automatic generation control (AGC) on Types C and D generating units, and the provision for the connection of nuclear plants. The PRC also requested changes (i) to the specified

---

45   A draft RPCC Articles of Association was prepared on 25 June 2014 but was never formalized by the GMS member countries. ENTSO-e Governance. https://www.entsoe.eu/about/inside-entsoe/governance/.

46   EU Commission Regulations. https://op.europa.eu/en/web/general-publications/publications.

fault levels, but it was agreed they could increase their internal fault levels and still comply with the RGC, and (ii) the reactive capacity of wind power plants, but this was rejected on the basis of EU experience. Given that the PRC is a world leader in wind technology, the second issue may need to be revisited when a new code for wind farms is developed.

Most of the 25 comments from Viet Nam resulted in clarifications of the basis for load frequency control setting based on what was agreed on in the GMS performance standards. Viet Nam also expressed concerns about the specifications for short circuit levels and clearing times being less stringent than the Viet Nam national code. Viet Nam indicated it may use a higher standard than the agreed performance standards (Section 5) for its hydro and thermal power station frequency control ranges which would anyway be compliant with the GMS. Some of the responses indicate that Viet Nam would be submitting further information about the Type D generation settings and testing, particularly where they are connected directly to busbars above 300 kV. Viet Nam also required more clarification regarding the requirements for HVDC back-to-back facilities and the studies necessary for evaluating the dynamic performance of interconnections between grids. It is understood that Viet Nam has considered these issues in its recent NGC review.

The gap analysis for the RGC CC notes that most aspects of the NGCs for the ASEAN members are generally in compliance with the CC, except for a few omissions and clarifications to be corrected and some modifications that may be necessary. These relate to the "necessary updates" particularly to the ability of demand facilities to remain connected after a fault and the specification of necessary tripping times. In the review of the national connection codes, it has been established that additional investment will be required to upgrade the governor and control systems of some of the existing generators to ensure they comply with primary and secondary frequency control requirements. There is also an issue in all three countries with some generators having inadequate voltage control due to their inability to generate or absorb their full share of reactive power. However, these situations can be mitigated by installing suitable FACTS devices as described in Section 3.

The investigation found that some of the newer Lao PDR IPPs connected to the EGAT system (e.g., Nam Theun 2 and Houay Ho) have settings that are overly sensitive to frequency fluctuations. This would make them liable to disconnect from the system during prolonged frequency excursions within the allowable limits of the RGC. Unless these settings can be corrected, it may be necessary for the TSO to issue special exemptions to enable each IPP to operate under restricted circumstances until the situation can be resolved.

The Operation Codes

The GMS Operation Code (OC) contains details of high-level operational procedures such as demand control, operational planning, and data provision. It has three subcodes totaling 202 pages. While the respective GMS OC subcodes do not cite the source of their information, it is evident they closely follow the guidelines and format

Table 10: Status of ENTSO-e Operation Codes Adopted in the GMS RGC

Operations Code	EU Commission Regulation	Internet Reference
Operations Security	EU 2017/2196	ENTSO-e. Operational Security. https://www.entsoe.eu/network_codes/os/.
Operational Planning and Scheduling	Pending	ENTSO-e. Operational Planning and Scheduling. https://www.entsoe.eu/network_codes/opss/.
LFC and Reserves	Pending	ENTSO-e. Load Frequency Control & Reserves. https://www.entsoe.eu/network_codes/load/.
Emergency and Restoration	EU 2017/2196	ENTSO-e. 2017. Network Codes Published. https://www.entsoe.eu/news/2017/11/28/network-codes-published/.

ENTSO-e = European Network of Transmission System Operators for Electricity, EU = European Union, GMS = Greater Mekong Subregion, LFC = load frequency control, RGC = regional grid code.

Source: ENTSO-e. System Operations. https://www.entsoe.eu/network_codes/sys-ops/

of the respective ENTSO-e codes, two of which are still pending in EU law as shown in Table 10.

The GMS OC subcodes focus on TSO rules for managing operational security, planning and scheduling, load frequency and reserves control, and energy and restorations when power systems are synchronously interconnected. They also apply to a TSO supplying power asynchronously via HVDC that can impact frequency stability or operational security. The GMS OC deals with the criteria and procedures that will be required to facilitate efficient, safe, reliable, and coordinated system operation of the GMS incorporated in the following subcodes. Each subcode is, in effect, a stand-alone document that duplicates many general provisions relating to regulatory aspects and approvals, recovery of costs, confidentiality obligations, and agreements with TSOs not bound by the RGC. The codes provide a list of related terms used to monitor and manage power flows within the interconnected systems and an explanation of the regulatory and confidentiality aspects of reporting relating to approvals, cost recovery, consultation, and coordination. Unlike the other GMS RGC codes, each subcode of the OC also includes a clause relating to its entry into force requiring a signoff by the TSOs and a date when it is applicable.

There are many issues identified in the gap analysis that need to be addressed by the TSOs before the RGC can come into force. It was evident that most of the questions raised during the discussion of the relevant subcodes arose because of the inadequacy of the existing operating codes in the NGCs. It appears that many of the respective operational rules are held internally by the TSOs but need to be aggregated and reorganized in the same format as the RGC so they can be made available to other stakeholders interested in power trading.

It is also apparent that more discussion is needed to agree on common national standards that comply with the RGC and enable TSO plans to be made more transparent to market players. Viet Nam has identified this area as a significant matter of concern, recognizing that operations codes must be transparent to a larger number of stakeholders including power plant operators, TSO/DSOs, power system design engineers, and grid infrastructure investors. Oversight of these activities will require the RPCC to be established as soon as possible to facilitate coordination between TSOs and ensure consistency in their planning and scheduling as soon as the RGC comes into force. To move forward with the GMS power trading plans, the RPTCC will need to prioritize its activities to ensure the TSOs work closely to bring their respective operation codes into line with the provisions in the GMS RGC.

Operational Security Code

The GMS Operational Security Code (OS) subcode defines the general requirements and principles for transmission systems applicable to all TSOs, DSOs, and significant grid users in Normal and Alert System States. It names the provisions in relation to the emergency state, blackout state, and restoration. The code does not apply to member states that are running asynchronously or temporarily disconnected from a synchronous area. The OS covers the operational security requirements in terms of system states, frequency and voltage control, short circuit and power flow management, contingency planning protection, and dynamic stability. It also sets out the requirements for data exchange and responsibilities of RPCC and TSOs in using the interconnected transmission system. As noted in the equivalent ENTSO-e OS code:

> This code is the first one in the field of system operation and serves as the 'umbrella' code for all the system operation network codes. It sets the overall principles for system operation, reflects on the common issues in the Network Codes for Load Frequency Control and Reserves, for Operational Planning and Scheduling and identifies the key issues to be dealt with in detail in the Network Code for Emergency and Restoration according to the requirements in the Framework Guidelines on Electricity System Operation.

The GMS OS code includes 19 comments from the PRC, Thailand, and Viet Nam, mostly seeking clarification or requesting minor changes that were accepted. Issues were raised by the PRC and Thailand about the definition of emergency states, contingency and security analysis, and requiring feedback from TSOs. The PRC raised concerns about the long time (30 minutes) needed to declare an emergency, and the long distances between member connections. The PRC also questioned the basis for specifying the data exchange requirements but agreed that more simulation studies had to be done by the RPCC to calibrate the analytical models to better evaluate contingency situations. It was agreed with Thailand that further work needs to be done by the RPCC to decide the parameters that should be monitored for the analysis of contingencies. Viet Nam requested a clearer definition of N-2 conditions and protection requirements that were agreed to be incorporated into the code. Viet Nam also raised issues about the frequency of exchanging high-quality data for analysis and agreed that the nature of data needed would be jointly analyzed by the respective TSOs.

The gap analysis has 25 recommendations for upgrading the respective NGCs to specify operational security limits for each line in the network for monitoring system states, LFC requirements, and active power reserves. Most of these recommendations are for additions to the existing codes considered critical to achieving regional system stability. In many cases, the information required by the TSOs to manage the national systems is not specified or ambiguous. At least 50% of the 30 main subheadings of the codes have missing clauses largely relating to the inadequate specifications of operational security limits as required for security analysis and contingency planning. Provisions also need to be added to the NGCs to cover organizational responsibilities, protection set points, and dynamic stability assessments. These will require several additions to cover the measurement and monitoring procedures agreed upon when frequency excursions exceed the 49–51 Hz limits.

Operational Planning and Scheduling Code

The Operational Planning and Scheduling Code (OPS) defines the minimum operational planning and scheduling requirements for ensuring coherent and coordinated operational planning processes of the synchronous areas applicable to all significant grid users, all TSOs, and all DSOs. The subcode describes the data needed for security analysis, the process for undertaking the analysis in operational planning, how outages should be coordinated, the need for forecasting, the role of ancillary services, how generating scheduling should be performed, and the operational planning data requirements and performance indicators. The purpose of this code is to (i) decide common time horizons, methodologies, and principles to carry out coordinated operational security analysis, and adequacy analysis to maintain operational security and support the efficient functioning of the GMS electricity market; and (ii) determine conditions to coordinate availability plans allowing work required on relevant assets.

The OPS subcode incorporates 16 comments from the PRC, Thailand, and Viet Nam, mostly seeking clarification or requesting minor changes that were accepted. Comments by the PRC and Thailand requested detail and case studies that are more appropriately included in a separate guidelines document for RPCC/TSO use. The PRC requested more explanation about the method for coordination between TSOs to analyze operational security analysis. There were also issues raised about the confidentiality of planning information and the need for sharing plans between TSOs. It was agreed that further work would be necessary to elaborate on these questions in a separate agreement between the TSOs and the RPCC. This would cover the number and types of scenarios to be developed, the models to be used and examples of the type of analysis to be completed. Some of these provisions should be incorporated in the balancing and interchanging provisions incorporated in the market code. Thailand sought clarification on the frequency of operations security analysis and the application of FACTS devices. Responses to other comments were more appropriately covered in a separate section of the market code. Changes requested by Viet Nam about outage coordination were accepted and the code changed accordingly.

The gap analysis found most of the 20 or so specific provisions in the RGC need "to be added" to the national grid codes. The provisions would require the TSOs to codify their involvement in the coordination of operational planning, outage coordination, adequacy, ancillary services, scheduling, and data environment. The existing planning elements require details of grid modeling for hour, day, month, and year planning along with a statement of the method used. Likewise, outage coordination planning needs to be defined by region and published to enable stakeholders to make their own assessment of the risks of power trading. The information along with the opportunities for stakeholders to supply ancillary services and generation support need to be made available to the RPCC for dissemination to market players.

Load Frequency Control and Reserves Code

The GMS Load Frequency Control and Reserves Code (LFCR) subcode defines the minimal requirements and principles for load frequency control and reserves applicable to all TSOs, Reserve Connecting DSOs, and Reserve Providers needed to adhere to the Synchronous Area Block operation agreements. This 64-page subcode aims at (i) achieving and keeping a satisfactory level of system frequency quality and efficient use of the power systems and resources; (ii) ensuring coherent and coordinated behavior of the transmission systems and power systems in real-time operation; (iii) deciding common requirements and principles for frequency containment reserves (FCR), frequency restoration reserves (FRR), and restoration reserves (RR); and (iv) determining common requirements for cross-border exchange, sharing, activation, and sizing of reserves.

The LFCR establishes costing methods for the operational and monitoring agreements including imbalance netting, cross-border activation, and exchange sharing. The subcode defines frequency standards, data collection and monitoring of load and generator ramping behavior, and mitigation. It also covers the aspects of monitoring sharing and exchanges of reserves including cooperation between TSOs and DSOs when the codes are in force.

Despite the length and complexity of the LFCR, there were no comments attached to this document. The gap assessment found only seven specific provisions that need "to be added" in the NGCs relating to synchronous area operational agreements, imbalance netting process, cross-border activation agreements, sharing agreements, and real-time data exchange. In particular, the TSOs need to establish load frequency control blocks and area operational agreements for imbalance netting, cross-border activation, sharing of reserves, and real-time data exchange.

A key concern is the lack of AGC facilities on many generators, where settings are currently controlled on a manual basis. It is also important that the planned new control centers in these countries are provided with control and communications facilities to ensure the national AGC system follows the RGC. In this respect, the investigating report provides a considerable explanation of the issues that need to be addressed along with proposed actions to be taken to rectify the problems.

The GMS LFCR is, however, a substantive and important code and will need to be reviewed in more detail during the process of deciding on a strategy for synchronization of the GMS grids. In particular, the provisions under Section 9 Exchange and Sharing of Reserves will be important for a grid-to-grid project using HVDC prior to any attempt to synchronize grids.

Emergency and Restoration Code

The Emergency and Restoration Code (E&M) defines the requirements and principles applicable to an emergency state, blackout state, and restoration to TSOs, DSOs, significant grid users, defense service providers, restoration service providers, market participants, and any third party that has a role in the market for the efficient use of the power system and resources. The subcode sets out the requirements for data exchange and responsibilities of RPCC and the TSOs in running the interconnected transmission system. It lays down minimum requirements for (i) the management of emergency, blackout, and restoration states; (ii) the coordination of the GMS system operation in a common and coherent way; (iii) simulations and tests for reliable, efficient, and fast restoration; and (iv) the tools and facilities needed for the purpose of reliable, efficient, and fast restoration.

The code received six comments from the PRC and Thailand seeking clarification or requesting minor changes that were accepted. The PRC had concerns with the application of under-frequency load shedding and the role of the TSO as a frequency leader and the provision for compliance testing and periodic review of the defense plan. Thailand had similar concerns relating to the steps necessary to deal with under-frequency control. It was agreed that further work would be necessary to decide if the proposals by the PRC and Thailand were workable. The gap assessment found two items that need "to be added" in the NGCs relating to a system defense plan and a restoration plan to return the system to its normal state as soon as possible.

The Market Codes

The 20-page GMS Market Code (MC) (not cited in the preamble) has a set of operational requirements for the GMS market incorporating the specific subcodes for capacity allocation and congestion management, forward capacity allocation, and electricity balancing. These codified guidelines are an integral part of the market code family. The two subcodes set out non-discriminatory rules for access conditions to the network for cross-border exchanges in electricity and rules on capacity allocation and congestion management for interconnections and transmission systems affecting cross-border electricity flows.

These subcodes are based on codes and policies stated in EU Commission Regulations 2015/1222, 2016/1719, ENTSO-e Policy 4 v2.4 and a draft Balancing Code, accessed November 2017 and adapted for the GMS Interconnected Network. The progress made by the European TSOs in implementing these regulations is documented in two reports which have now been published by ENTSO-e.[47] The ADB knowledge product on *Regulatory and Pricing Measures to Facilitate Power Trade* makes proposals on how the respective cost and benefits of grid-to-grid power trading can be distributed to the parties involved in each transaction. It describes open access proposals, methodologies for wheeling charges, short-term bilateral trading measures, and

---

[47] ENTSO-e. 2019. *ENTSO-e Market Report 2019.* https://eepublicdownloads.blob.core.windows.net/public-cdn-container/clean-documents/mc-documents/190814_ENTSO-E%20Market%20Report%202019.pdf.

balancing mechanisms based on the provisions in the market codes. It also provides several international case studies, examples of regional balancing arrangements, and a detailed method and application for wheeling charges. It notes that

> ...many of these requirements are encapsulated in the concept of open access, to enable generators to use the transmission assets of their host national power utility to wheel power to the national border for export purposes. Open access to transmission networks is a key requirement for independent electricity trading parties (i.e., businesses that are separate from the owners of national transmission networks) to trade electricity either with (a) customers within the national network directly, or (b) power utilities or large consumers connected to the networks of neighboring countries

Capacity Allocation and Congestion Management

This subcode sets out non-discriminatory rules for access conditions to the network for cross-border exchanges in electricity and rules on capacity allocation, forward capacity allocation, and congestion management for interconnections and transmission systems affecting cross-border electricity flows. It proposes that RPCC establish a common grid model including estimates on generation, load, and network status for each hour. The available capacity should be calculated according to the flow-based calculation method, a method that considers that electricity can flow via different paths and optimizes the available capacity in highly interdependent grids. Each TSO in the interconnected system is required to prepare an individual grid model of its system and send it to RPCC for merging into a common grid model. The individual grid models should include information on generation and load units.

The GMS Capacity Allocation and Congestion Management (CACM) describes the concepts, capacity calculation methodology, and time frames for determining physical transmission rights and financial transmission rights. It sets out the methodology for providing generation and load data and how this will be allocated to the transmission capacity calculations that consider reliability margins and operational security limits in determining net transfer capacity limits. It also sets out the rules and dispute mechanisms for computation and publication of long-term capacity. This subcode covers day-ahead and intra-day markets, and the available cross-border capacity needs to be calculated in a coordinated manner by RPCC. For this purpose, RPCC should establish a common grid model including estimates on generation, load, and network status for each hour.

There were four comments from Thailand. One proposed a change to recognize the GMS dry and rainy seasons and to make minor corrections to the formulas that were accepted. A proposal to adopt the North American Electric Reliability Corporation reliability standard was rejected to prevent legal disputes. The gap assessment identified three items that need "to be added" in the NGCs relating to hourly transmission allocations, physical allocation of transmission rights, and the calculation of cross zonal capacity.

Electricity Balancing

This subcode establishes a GMS-wide set of technical, operational and market rules to govern the functioning of electricity balancing arrangements. It sets out rules for the procurement of balancing capacity, the activation of balancing energy, and the financial settlement of balancing energy. Its main objectives are (i) enhancing the efficiency of balancing GMS and national balancing arrangements and (ii) integrating balancing arrangements and promoting the possibilities for exchanges of balancing services while contributing to operational security. The subcode explains the role of the TSOs as service providers and the responsibility of the RPCC for developing a platform for TSO–TSO exchanges. The RPCC sets the rules for determining the balancing mechanism and the basis for calculation in the absence of a balancing market. It defines how imbalances should be settled and reported according to the causes.

Thailand made only four comments on the code. It requested provision for wet or dry seasons (as opposed to summer or winter in ENTSO-e) and questioned why United States standards should not be applied. However, it was agreed to adopt the ENTSO-e provisions without performance criteria. The gap analysis identified two items that need "to be added" in the NGCs. These relate to the obligation for TSOs to make bids for surplus balancing and to invite other TSOs to offer balancing services when it is economic to do so.

The Metering Code

The 13-page GMS Metering Code is a general specification that defines the metering types and functions for use at each point of exchange between grid control areas. It specifies the minimum technical, design, and operational criteria to be complied with for metering each point of interchange of energy between control areas, TSOs and other trading parties. The code is not concerned with metering connection points identified in the connection codes at which (i) the power generating plant is connected to a transmission system or distribution network, (ii) the demand facility or the distribution network is connected to a transmission network, or (iii) the closed distribution network providing demand-side response is connected to the wider distribution network. These commercial metering systems are subject to NGC regulations and PPAs.

The GMS Metering Code refers to the IEC standards for functional, mechanical, electrical, and marking requirements, test methods, and test conditions, including immunity to external influences covering electromagnetic and climatic environments.[48] It describes the meter placement, compliance of metering hardware according to standards in terms of accuracy levels, accessibility to meters, responsibility for meters, and maintenance. The GMS metering code also specifies the associated data collection equipment and the related metering procedures

---

[48]   IEC 65052-11:2020; IEC 60044 (Parts 2–5). https://www.smart-energy.com/industry-sectors/policy-regulation/iec-standards-for-electricity-metering/.

required for the operation of the interconnected transmission system and sets out provisions relating to (i) main and check metering installations used for the measurement of active and reactive energy; (ii) the collection of metering data; (iii) the provision, installation, and maintenance of equipment; (iv) the accuracy of all equipment used in the process of electricity metering; (v) testing procedures to be adhered to; (vi) storage requirements for metering data; (vii) competencies and standards of performance; and (viii) the relationship of entities involved in the electricity metering industry.

Although there were no comments attached to this document, its scope will need to be expanded to cover provisions to facilitate energy data exchange and interoperability to develop a single homogenous model or set of rules to fit all GMS members.[49]

The System Operator Training Code

The 12-page GMS System Operator Training Code (SOTC) sets out the responsibilities and the minimum acceptable requirements for the development and implementation of system operator training and authorization programs. The code defines a standard framework for operational training to provide reasonable assurance that dispatchers have and keep the knowledge and skills to always run the power system in a safe and reliable manner under all conditions. It will ensure that system operators throughout the GMS are provided with continuous and coordinated operational training to promote the reliability and security of the interconnected transmission system. This code was based on ENTSO-e Policy 8 Operator Training, accessed November 2017 and adapted for the GMS RGC.

The PRC made three comments regarding prerequisites for training that were accepted. These covered safety codes and re-examination of dispatchers if they have been absent for more than 6 months.

GMS Strategic Planning Document

The 11-page GMS Strategic Planning Document (SPD) specifies the minimum technical and design criteria, principles and procedures for medium- and long-term development of the GMS synchronously interconnected transmission systems along with the planning date required to be shared among members. It specifies the requirements for the interchange of information between the RPCC and individual TSOs.

This information is needed to enable the RPCC and TSOs to take due account of regional developments, new connection sites, or the modification of existing connection sites within the TSO's transmission network along with new or modified connections with external systems. The SPD code defines the reliability criteria,

---

[49]   EU-SysFlex. 2019. *European Level Legal Requirements to Energy Data Exchange.*

the requirements for cost-benefit analyses, and details of the planning process including provisions for a power balance statement and a transmission system capability statement. It also defines the power system modeling requirements in terms of responsibilities, modeling data, and confidentiality. There are no comments on the code and no mention of its requirements for compatibility with individual country planning codes.

6. Implementation of Power Trading

Moving toward a Regional Power Market

Progress toward achieving open access GMS power trading has been challenging, particularly regarding the aim to achieve full synchronization of the GMS grids. To some extent, this is due to the TSOs' conflicting objectives of maintaining the security and reliability of their electricity supplies against the national objective of reducing the cost of unsubsidized power to consumers. While this issue may be resolved by enforcing the new RGC, there is still much to do to prepare for the commencement of grid-to-grid power trading in the GMS as envisaged under the 2012 Road Map. Hopefully, a tentative start can be made soon with both Stage 2 (implementing the RGC) and Stage 3 (building the interconnections), although it is evident that the final Stage 4 is some way off. This was envisaged as the time "most of the GMS countries have moved to multiple sellers–buyers regulatory frameworks, so a wholly competitive regional market can be implemented." [50]

During the next few years, ADB's continuing leadership and the flexibility of its technical assistance program along with the support of its development partners will be essential. Continued participation by international financing agencies will help the RPTCC working groups focus on implementing the minimum required attributes of the RGC and establishing the institutional capability to enforce it. In this respect, two recent International Energy Agency (IEA) publications (Appendix 3) have proposed valuable recommendations for taking the next steps to establish the institutional arrangements to support the development of a GMS/ASEAN regional power market. In its most recent study, the IEA report covers Thailand and suggests that there are issues in the Thai grid codes used by TSOs and DSOs that could be improved, not only to accommodate the increasing amount of wind and solar PV but also to seek consistency and alignment across all the Thai grid codes.

Until the RPCC is established and permanently staffed it will be difficult to achieve a fully synchronized GMS transmission network before 2030. In the meantime, the RPTCC needs to be empowered to undertake the outstanding tasks relating to the promotion of efficient electricity markets, information exchange between single buyers and regulators, and a statement of opportunity for investors. Future focus areas could also include the following considerations:

(i)　Interface management with nested and fragmented control areas that exist in the Lao PDR and Cambodia.

50　ADB. 2016. *Greater Mekong Subregion Energy Sector Assessment, Strategy, and Road Map*. Manila. https://www.adb.org/sites/default/files/institutional-document/188878/gms-energy-asr.pdf.

(ii)   Digital infrastructure, cyber security, and interoperable information and communication systems.

(iii)   Energy and capacity procurement with physical and financial market structures.

(iv)   Planning challenges of geographically concentrated large-scale renewables.

(v)   Essential reliability services from renewables and customer-owned resources along with voltage control and black start ancillary services.

(vi)   Risk-based probabilistic transmission planning processes and criteria along with nonwire alternatives.

(vii) DSO participation in power trading.

(viii) Offshore wind and hybrid renewables (wind–solar, wind–solar–storage, etc.).

(ix)   Inertia monitoring and assessment.

The working groups need to continue to advise GMS policymakers and regulators on the steps necessary to approve wheeling charges, formally adopt the RGC as a basis for harmonization of the NGCs, and commence the design of contractual structures and technical standards for new interconnections. At the same time, the RPTCC will need to widen stakeholder participation in the continuing development of the RGC to include DSOs, IPPs, large industries, and equipment manufacturers. Such cooperation could be initiated through a common information facility that promotes transparent open access and invites submissions from stakeholders for proposals on how to reflect the special characteristics of the GMS.

As a sign of the significant level of activity required by the RPTCC to support and develop the GMS RGC, it may be useful to review the ENTSO-e Work Programs from 2015–2020 as a proxy for considering what must be done to update the RGC.[51] Each year, the work program describes the strategic planning necessary along with the scope of the continuous committee activities planned for the year ahead.

Road Map for the Implementation of the Regional Grid Code

The technical work required to complete the implementation phase of the RGC along with the associated development of interconnections is summarized in Section 5. This work program identifies specific tasks to be undertaken under the headings (i) a new Road Map for implementation of the GMS RGC, (ii) operationalization of the GMS synchronous areas, (iii) design of RPCC communication and data management system, and (iv) metering organization and architecture (using existing facilities as much as possible). Two additional planning tasks should include (v) providing guidance to the RPTCC for the implementation of the regional transmission MHI Master Plan, taking into consideration the current plans by EDL-T to build a hybrid 500 kV/HVDC backbone grid (Section 2), and

---

51   ENTSO-e Annual Work Programme. https://www.entsoe.eu/publications/general-publications/awp.

(vi) assessing how new interconnection projects can be prioritized for investment that will accelerate the implementation of a fully synchronized grid. From a conceptual viewpoint, the work program for establishing the power market is shown in Figure 13.[52]

These tasks are designed to identify the constraints to increasing renewable energy penetration in the GMS and establish the key technical priorities that need to be addressed at a regional level. The WGPO will need to consider the commercial arrangements that need attention in the development of the RGC to support a variety of market platforms, a balancing market, and defined market rules and market surveillance arrangements. This will result in developing a framework for regulatory oversight to achieve a balance between national and regional regulators. The WGRI will also need to review the development of traditional hydro, coal, gas, and geothermal projects in Asia, noting that most wind and solar projects are more likely to be implemented by private sector operators. In this respect, private sector participation in the GMS countries lags behind Singapore and Malaysia where VRE projects were more bankable and more easily integrated within the power system.

Figure 13: Priority Rules for the Harmonization of GMS Grids

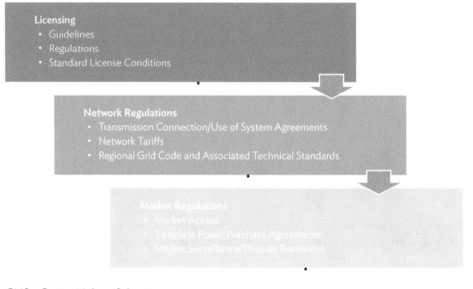

Licensing
- Guidelines
- Regulations
- Standard License Conditions

Network Regulations
- Transmission Connection/Use of System Agreements
- Network Tariffs
- Regional Grid Code and Associated Technical Standards

Market Regulations
- Market Access
- Template Power Purchase Agreements
- Market Surveillance/Dispute Resolution

GMS = Greater Mekong Subregion.
Source: D. T. Bui and J. Hedgecock. 2020. RPTCC WGRI Meeting Paper. 27 October.

---

[52] During RPTCC 27 in October 2020, two papers were presented by World Bank and ADB consultants titled Emerging Trends in GMS Power Sector Development covering (i) integration of renewable energy, and (ii) private sector engagement. The latter paper includes a summary of the country discussions that examined the issues in each of the GMS countries that should be considered in the WGRI work plans to encourage more inward investment in GMS countries.

Institutional Issues

Because of the impasse over where the RPCC should be located, it may be appropriate to consider leasing a temporary RPTCC administration center in a convenient central location. The Lao PDR has considerable advantages in enabling cross-border access to its hydropower storage capability along with the ability to develop secure cross-border fiber-optic communication links to all the other GMS countries. If it was possible, such a center could be co-located in or near the planned headquarters of the new EDT-L Transmission company. The main functions of the center would be to address the key issues of how to direct and coordinate the work by the working groups while also liaising with the national regulators and other stakeholders. It should be a priority for the RPTCC to make a presentation to the regulators to explain the status of the RGC and seek advice on how it should be developed to become enforceable.

The establishment of an administrative center would enable the RPTCC to work full time on the many procedural actions that need to be developed to bring national operational processes and procedures and rules into compliance with the GMS Grid Code. These include procedures for security analysis, outage scheduling, a national grid model, a load frequency control methodology, restoration plans, and balancing mechanisms. It would also enable the RPTCC to propose technical and cost remedial measures to bring non-conforming power plants into compliance along with implementation plans to develop mechanisms for monitoring the GMS TSOs.

TSOs have often expressed concerns about sharing technical and commercially sensitive data. The limited access to such technical and commercial data, and the use of common software were often cited as reasons for delays in key studies of cross-border projects.[53] There also appears to be a lack of a suitable institutional memory bank for the products of RGC development processes, with no centralized database where past information such as country and working group reports can be archived. In this respect, the RPTCC should prepare a standard pro forma reporting system with an emphasis on power trading results.

To address these issues, there is an urgent need for a cloud storage facility where TSOs can store and exchange such information under strict rules of confidentiality and cybersecurity. The working groups will need to develop standard database protocols for maintaining power system data and procure new software tools for planning and analysis of power system network design using emerging new smart grid technologies to optimize the use of interconnections. The tools would include geographic information system mapping for asset management, communications systems for monitoring to improve use, and modern maintenance practices to increase asset longevity. These tools could be obtained and licensed for use by the GMS TSOs with access to a protected but shared GMS power system database.[54]

---

[53]   As noted by Viet Nam (RPCC 27), it should be possible to negotiate with software suppliers for member countries to have access to such software through the RPCC after it is established.

[54]   The country reports to the RPTCC meetings are largely focused on domestic results with very little information on imports and exports within the region.

Regional Power Trade Coordinating Committee Working Group Programs

As required under Section 5 of the Governance Code, a Grid Code Secretariat along with a Review Panel will need to be established to consider how the RGC can be enforced as part of the process of approving new investments. The RGC Secretariat could initially operate virtually from within the TSO offices and be coordinated by an RPTCC administrator. This will enable the working groups to review and continuously update progress in addressing and costing the measures recommended in the gap assessment report, including updating respective NGCs for generation, transmission, and distribution. The RGC Review Panel would develop mechanisms for compliance monitoring at a regional level and obtain agreements with the TSOs for ensuring their commitments.

As recommended in Appendix 7 (Task 6), the working groups will also need to organize their activities to bring more focus to establishing separate groups for the following:

(i)   Operationalization of GMS synchronous areas, including operational agreements, methodologies, processes, and procedures, will require the formulation of multiparty agreements and common methodologies for securely coordinating operations in each synchronous zone.

(ii)   Establishment of a load frequency control group that will complete the assessment of the capability within each TSO to manage frequency control within the limits specified in the GMS RGC.

(iii) Development of regional control and communications systems and associated metering architecture to be used as a basis for securing financing for the development of a single turnkey project.

A major concern is how to initiate open access grid-to-grid power market trading before the GMS countries are fully synchronized. In the near term, it appears unlikely that TSOs will agree to a fully synchronized 500 kV regional grid operation and will need to gain confidence in monitoring and controlling the impacts of interconnections with each other. In this respect, the deployment of grid-to-grid HVDC-VSC interconnections with torque synchronizing capability may offer a way forward.[55] The PRC has already designed such facilities that can operate in parallel with future HVAC lines that are unlikely to become redundant after full GMS synchronism is achieved.

---

[55]   J. C. G. Torres et al. 2020. Transient Stability of Power Systems With Embedded VSC–HVDC Links: Stability Margins Analysis and Control. *IET Generation, Transmission & Distribution.* 14 (17). September. pp. 3377–3388. https://doi.org/10.1049/iet-gtd.2019.1074.

Plans for the Working Group
on Regulatory Issues

Pilot Wheeling Project

A pilot wheeling initiative was proposed by ASEAN members in 2013 to transfer 100 MW from the Lao PDR to Singapore.[56] Despite having agreed on commercial arrangements including signing an energy purchase and wheeling agreement in 2017, the project stalled in its final phase of having to meet the obligations of the Singapore power market. Since the relevant part of the Lao PDR exporting grid and Thailand are already synchronized, this project would be, in essence, complying with the Thailand NGC with no need to await formal approval from the RGC. Moreover, the project would take advantage of the controllability of the existing HVDC back-to-back connection between Malaysia and Thailand and provide a useful model for other HVDC power trading arrangements in the GMS. It would enable the GMS countries to gain experience in compliance with the Malaysian and Singapore market codes while enabling Thailand to closely monitor the impacts of the power flow across both borders. The re-establishment of the pilot grid-to-grid power trading project will help uncover many issues that still need to be addressed before GMS power trading can begin in earnest. One objective would be to identify practical problems in legal and regulatory structures, make proposals for fixing them, and prepare regulatory guidelines to assist in their implementation. This would help trial the use of the recommended methodologies for determining wheeling charges based on the extent of the transmission system asset base and the agreed weighted average cost of capital. The main objective of the project would be to identify how costs of power trading across an intermediate grid would be assessed and shared.

Independent Power Producer Stakeholders

The commercial interests of the existing IPP investors must be considered so they too can enter the power market on a mutually satisfactory basis. Most private companies in the GMS market are continuing to invest in long-term generation projects and are increasingly locked into long-term PPAs that do not cater to regional power trading. Some of the longer-running PPAs may now be open to renegotiation shortly before their terms expire, with the new contracts required to be compliant with RGC. In the meantime, the WGRI needs to investigate ways in which the power trading regime can be made to align with existing IPP commercial and technical conditions in consultation with regional regulators.

Integration of Variable Renewable Energy

Most GMS power grids have been designed for traditional fossil fuel resources and not renewables. Substantial investments in the existing power systems are needed to

---

[56]   ADB. 2020. *Harmonizing Power Systems in the Greater Mekong Subregion: Regulatory and Pricing Measures to Facilitate Trade.* Manila. Appendix 3 http://dx.doi.org/10.22617/TCS200070.

enable the integration of renewables and the rapid increase in demand resulting from the electrification of transport and heat. Given the currently low level of renewable technology penetration in the regional energy mix, the creation of an electricity storage market will be a major factor in supporting the eventual adoption of a large percentage of distributed renewables in the GMS countries. This would help bolster renewable electricity sources that could be used during peak demand days during the warmer dry season and eliminate the need for some of the additional reserve capacity in the long term.

A Thai developer is already constructing a 1.8 MW battery electric storage system (BESS) facility associated with a 10 MW wind development project and will soon be in a position to demonstrate how this facility contributes to its operations. An even larger BESS project is being considered by a private developer as part of a 200 MW industrial microgrid.[57] These projects could provide the RPTCC with an opportunity to study the potential for large grid battery or pumped hydropower storage and consider the price points at which BESS technology would be worthwhile in other locations.

Strategic Planning of Interconnections

The RPTCC will need to review the strategic plans for interconnection investments to ensure the contracting agreements meet the objectives of the proponents and serve the common interest of developing grid-to-grid power trading facilities. Some projects could be financed by private investors using special purpose vehicles to manage interconnections on behalf of the respective grids. Others, particularly where there is a wider strategic objective such as accelerating the GMS synchronization process, might require the support of an international financial institution backed by a sovereign guarantee to ensure transparent and equal treatment of the beneficiaries. The plans for a joint venture between the Lao PDR and the PRC to establish a hybrid HVDC/500 kV grid in the Lao PDR is a case in point. Such an agreement will need to consider the following concerns:

(i)   Ensure the fair distribution of economic, social, and other benefits and costs among the nations involved in an interconnection as well as stakeholder groups within nations.

(ii)   Ensure that the direct and avoided costs of an interconnection are specified as accurately as possible, preferably within the context of comprehensive long-term power system planning. This includes assessments of environmental costs and benefits.

(iii) Emphasize transparency in all negotiations related to grid interconnections, including allowing all stakeholders access to all relevant materials. Include all potentially affected stakeholders in the early stages of project formulation and continue to solicit the input of all contracting parties on key decisions throughout the project.

---

57   A. Colthorpe. 2021. Hitachi ABB Power Grids' Battery Storage To Be Used at 214MW Industrial Microgrid in Thailand. *Energy Storage News.* 14 May. https://www.energy-storage.news/hitachi-abb-power-grids-battery-storage-to-be-used-at-214mw-industrial-microgrid-in-thailand/.

(vi)   Establish clear needs and protocols for collecting and distributing quantitative data and other information needed for project design, as well as for the accurate estimation of project costs and benefits. Establish clear legal and administrative authority over all aspects of the design, construction, and operation of the grid interconnection.

(v)   Locate new power lines in existing transmission or transport corridors as much as possible. Implement capacity building to allow different social stakeholder groups to meaningfully participate in investigating and deciding on grid interconnection options and planning for grid interconnection construction and operation.

Plans for the Working Group Planning and Operations

Synchronization of GMS Grids

The RPTCC requires a strategy for synchronizing all the GMS grids within a decade, preferably by prioritization of strategic elements of the recommended interconnections listed in Table 4. The WGPO needs to examine alternative ways in which synchronization could be achieved using technologies such as VFT border interconnections (e.g., between the Lao PDR domestic and international systems) and for longer cross-border interconnections using HVDC-VSC facilities fitted with torque synchronizing capabilities.

Because the frequency of the respective power systems will always be subject to oscillations, regional power systems will need robust LFC operations to ensure they comply with stability limits. There are many ways frequency control is administered and it may be possible to structure a new LFC system that can be used across borders. This would include a strategy to enable greater uptake of VRE where uncertainties of active power production are significantly increased and by active power fluctuation in combination with demand stochasticity. Over the last decade, proven new mitigation technologies have emerged to support the increased penetration of low-cost VRE. In many cases, they are designed so that private sector investments can be efficiently integrated into the existing national power market.

Based on the findings of the investigations of NGC compliance with the RGC, it appears that Cambodia and the Lao PDR have inadequate capability to monitor and control the frequency of their respective domestic networks. For example, the frequency in Cambodia is controlled by Viet Nam and many of the Lao PDR IPPs are controlled from Thailand. These countries will need to invest in new governor and excitation systems to bring some of their existing generators into compliance with the new frequency standards.

On the other hand, NGCs in Viet Nam and Thailand are largely compliant with the RGC connection codes that also apply to the Lao PDR IPPs synchronized with both countries. The two countries could take the initiative to pave the way to a wider synchronization program by investigating ways to safely interconnect their respective 500 kV systems. They both have well-established NGCs and could, if they were directed to do so, cooperate with the new EDL-T transmission company to use the proposed 500 kV or HVDC network to facilitate the early implementation of grid-to-grid trading. They could start by establishing a commercial reserve sharing arrangement either through an HVDC-VSC interconnection or by synchronizing with each other through two or more interconnected 500 kV lines. As a starting point, the TSOs would need to assemble all the necessary technical power system data and carry out specially focused integrated power system stability and fault studies to investigate the impact of interconnections and the necessary precautions required under contingency situations.

Aligning NGCs with the RGC

The main concerns in the power grids of Cambodia, the Lao PDR, and Myanmar related to governor and excitation control settings on some generation plants that needed to comply with RGC frequency control standards. However, throughout the GMS, the NGCs will also require additions and modifications to operational and market codes to facilitate a transparent open market for power trading. As noted in Section 2, the review should also consider the NGCs of the independent distribution companies in Thailand and Viet Nam to ensure they can be accommodated in a power trading regime. To this end, the WGPO will need to consider the following recommendations:

(i)   Engage all future stakeholders in power trading to ensure that the codes can be implemented without putting system security at risk and responsibilities are fairly distributed between all actors.

(ii)   Put in place a predictable and reliable grid code revision process. This increases system reliability and security by coordinating changes as technology and operation practices develop, also facilitating the future planning of the system. It will also help continue to tailor the RGC to the changing characteristics of the GMS.

(iii) Establish regional initiatives and engage in international standardization processes may facilitate the development and implementation of grid codes by sharing experiences, deploying regional infrastructure for verification and certification processes, and harmonizing requirements resulting in cost reductions due to market scale for technology suppliers.

(iv)   Ensure that national and regional grid connection codes include appropriate requirements for VRE. The significant experience of the PRC and Viet Nam with VRE could be used as input for the code.

Enhancements to the RGC

There is also a need for further enhancements to the RGC to facilitate their acceptance by regulators. These include (i) providing more detail about the proposed governance structure that can be used to set up the RPCC and (ii) inclusion of the associated preambles to each subcode that will aid the respective regulators to clarify its purpose. Any changes should be signed off by all members of the WGPO and published on the RPTCC website for comments by stakeholders.

There is a continuing need for amendments to cater to TSO concerns about cybersecurity and data confidentiality along with accommodating new technologies already entering the GMS power market as described in Section 4. To cater to the expected rapid growth in VRE development, it will be necessary to review the standards for connecting wind generators and consider how the region's hydro reservoir storage can be used and how to mitigate problems such as inertia management. There will also be a requirement to include incorporation of codes catering to large-scale battery storage system investments that are currently being installed in Thailand, the increasing capability of new HVDC technologies to provide black starting, and frequency control (as recently developed in the PRC), along with advances in load frequency control options. The WGPO should also consider the emergence of new technologies such as variable frequency transformer technologies that could be used as a low-cost means of interconnecting smaller asynchronous systems.

Control, Metering, and Data Acquisition Systems

It is important that the RPTCC begin designing the regional communications system to allow for safe and secure exchange of data and information among GMS TSO market operators and the RPCC. This could be done by setting up a new RPTCC working group, with the representation of all relevant stakeholders, to define a series of recommendations about the minimum security requirements that should always be applied to smart grid devices, intercommunications, and interdependencies. Alternatively, the RPTCC could define guidelines that include the implementation of minimum security requirements for all devices, interconnections, and interdependencies for the deployment of new smart grids.

Centralized monitoring of regional power trading will also be more vulnerable to cybersecurity problems (Section 4) where an attacker can access one or more of the sophisticated GMS member SCADA systems through a direct connection to a weak link in a smaller system. The first step for ensuring the security of the intercommunications, interdependencies, and devices is to set up a series of minimum security requirements that must be met to protect overall grid connectivity. These security requirements and associated controls must be defined to achieve a minimum level of security that will ensure service continuity and resilience, both in public and private environments.

As an interim arrangement, centralized monitoring may be more easily achieved in a private intranet system. From a central location in the Lao PDR, five radial optic fiber links (possibly installed in new transmission ground wires) could be built in the Lao PDR and across national boundaries to the existing SCADA facilities in Kunming, Ha Noi, Bangkok, Yangon, and Phnom Penh. Later, alternative communication links could be built between centers to provide some diversity in case of link failures. The monitoring of decentralized regional generation will require sophisticated supervisory control and data acquisition (SCADA) and related grid operation tools. New functional modules will be needed for new system operation processes at RPCC, TSO, and DSO levels. To cope with the increased complexity of system operation, the existing SCADA systems must be enhanced to enable the management of a power system interconnecting with many fluctuating sources.

There will also be a need to establish a metering organization and architecture allowing for the settlement of power exchange, including meter equipment on tie-lines and a centralized data processing system. This will require an assessment of future needs in terms of power and energy-metered data on the interconnection tie-lines, meter specifications, the centralized data processing system to be developed, necessary backup and remote reading systems, and security of data exchange to preserve the confidentiality of the information protocols.

7. Conclusions and Recommendations

Summary and Conclusions

This report describes the activities of the GMS TSOs in reaching agreement on a range of complex technical and institutional issues prescribed in the RGC. It also outlines the history, nature, and application of grid codes around the world, noting that there are several similar activities underway in a dozen other regional groups of international power utilities. Over the last decade, new challenges have arisen in terms of rapidly changing VRE generation, FACTS, and BESS technologies to help combat climate change and unforeseen transmission system failure events. These will require continuous adaptation of the grid codes to enhance the capability of existing transmission networks and institutional changes to ensure the fair allocation of benefits from power trading.

Power planning studies have confirmed that significant savings can be achieved, particularly through reserve sharing, load shifting, and the joint exploitation of VRE and hydropower resources. The methodology for allocating the network costs associated with power trading among national stakeholders has been outlined in the complementary ADB knowledge product dealing with regulatory and pricing measures applicable to the GMS (footnote 56). This report builds on those guidelines, by addressing the technical and operational issues of interconnections along with the steps required to harmonize the use of existing NGCs within an overarching regional structure of the RGC. It considers the special characteristics of the region and the interests of its stakeholders along with global trends in technological development expected to drive the scope of future amendments to the GMS RGC.

The work to establish the RGC has been challenging for TSO members assigned to the RPTCC working groups, often with participants having to take time off their regular duties. The working groups have met more than 20 times to review a variety of related technical reports, investment proposals, and gap analysis reports that, in some cases, will require amendments to their own NGCs. In doing so, the TSO representatives have successfully negotiated a historic international agreement on a standardized set of codes used as a reference document for further development. In the process, the RPTCC working groups have built up confidence in working together, identified gaps that need to be addressed, and gained the support of the respective government ministries and regulators in pursuing the goal of promoting open access power trading.

There is still a long way to go to establish a transparent, open access regime of power trading in the GMS. However, without an RPCC permanently staffed by members assigned from the TSOs, it will be difficult to achieve a fully synchronized GMS transmission network before 2030. In the meantime, the existing RPTCC needs to be empowered to undertake several tasks relating to the promotion of efficient electricity markets, information exchange between single buyers and regulators, and statements of opportunity for investors. RPTCC working groups also need to be empowered to advise policymakers, develop and approve wheeling charges to enable them to continue to adapt the RGC, and advise on contractual structures and technical standards. Their main aim should be to increase stakeholder participation in providing an information facility to promote transparent open access that explains the rationale for the RGC and invites submissions for further amendments to reflect the special characteristics of the GMS.

Lessons Learned

The lessons learned from the process of developing the GMS RGC are applicable to other regions of ADB operations where power trading operations are being developed. The GMS process started in 2002 with the first RETA 6440 program of technical assistance, which perhaps had too much focus on economic and institutional studies with limited input from stakeholders. It began in earnest after the 2016 GMS Road Map (supported by RETA 8330) by focusing TSO stakeholders on the technical issues that needed to be addressed before trying to reach agreement on more complex aspects of a common power market. The lessons learned under this project confirmed similar recommendations made in the IRENA Grid Code reports and in a 2005 United Nations report.[58] Some of the potential strategies for reaching the necessary agreements to implement an interconnection project are as follows:

(i)   Train local and regional people in several general and specific professional areas, ranging from electricity transmission engineering and power flow modeling to finance, utility management, law, marketing, regulation, negotiation and arbitration, information systems and database development, planning and policy development, data collection, and environmental analysis. Training needs should be addressed through existing centers of expertise within the region.

(ii)   Maintain an institutional memory of past work by compiling and publishing information, including technical parameters of power loads and flows in national transmission systems, the status of national and regional regulatory, financial and legal systems, energy sector forecasts and planning results, demographic and social data, resource, hydrologic and environmental data, and data on the costs and performance of new energy and environmental technologies.

(iii)   Sponsor analytical activities that are hard for individual countries or private groups to sponsor, including power flow modeling of non-connected and

---

58   United Nations Department of Economic and Social Affairs. Multi-Dimensional Issues in International Electric Power Grid Interconnections: Conclusions and Recommendations for Follow-up. Section C. New York. 2005.

interconnected systems; analysis of market systems for power trading; analyses of economic, environmental, and social impacts (pre-, during, and post-project); and electric power sector and overall energy planning, including forecasts of demand for electricity and energy services.

(iv)   Provide support for engagement by means of events and processes where counterparts from different regions and countries, and even subnational stakeholders can communicate, work, and learn together. Such opportunities include (a) regional study groups on the technical, economic, legal and regulatory, political and social, and environmental aspects of interconnections to serve a particular area; and (b) national and regional stakeholder meetings regarding interconnection prospects in general and specific interconnection options and support for the intervention of stakeholder groups in interconnection planning processes, including capacity building, expertise, and project support.

Recommendations

In the absence of an established RPCC and despite the difficulty of working during COVID-19 lockdowns, it is essential that the RPTCC regain its momentum by taking the initiative to promote the use of the RGC as the modus operandi for a future power trading regime. It must ensure the respective governments, regulatory authorities, and stakeholders are well-informed of the progress in developing the RGC and seek their advice as to how to make it legally enforceable. In the meantime, the RPTCC should investigate the viability of establishing a temporary administration center (preferably in Vientiane) where it can begin its activities by taking over responsibility for the management of the RPTCC website to

(i)   Post the GMS RGC in its present form along with planned proposals for amendments that invite stakeholders to provide comments and proposals.
(ii)   Provide links to reports and documentation that explain the history and background of the RGC. This should include links to activities by other stakeholders, including international stakeholders.
(iii)   Provide links to GMS member countries' technical and regulatory websites containing relevant technical information and NGCs for public access.
(iv)   Provide links to news sources reporting relevant energy development activities in the GMS region.
(v)   Post opportunities for investors in cross-border transmission projects.
(vi)   Provide a separate password controlled web page for coordination of ongoing working group programs.

The WGRI needs to implement the recommendations in Section 8.3 of the ADB knowledge product for regulation and pricing along with the following proposals:

(i)   Address market issues identified in the 2019 gap assessment report relating to the aspects of the RGC dealing with emergency restoration, capacity allocation and congestion management, and electricity balancing.

(ii)   Implement a pilot grid–to–grid power trading project to develop and seek regulatory approval for short-term trading rules along with a balancing mechanism.

(iii) Review existing long-term power purchase agreements with a view to negotiating changes with more flexibility to support power trading.

(iv)   Consider the commercial advantages of developing a reserve sharing project between the two largest grids, Thailand and Viet Nam.

(v)   Prioritize interconnection investments identified in the MHI Master Plan study that offers the best way to achieve a staged synchronization program.

Similarly, the WGPO needs to develop a strategy for synchronization for implementation before 2030. This might consider alternatives such as promoting the synchronous interconnection of the ASEAN members of the GMS before considering how the subregion should be interconnected with the much larger grid in the southern part of the PRC. This might include an assessment of the following:

(i)   Complete a review of the gap assessments, sign off on any changes to the GMS RGC, and invite comments from stakeholders for further development.

(ii)   Implement the main work plan tasks described in Appendix 7.

(iii) Assess how many parallel HVAC links are required to minimize the prospect of synchronous separation due to adverse system incidents (such as recently experienced in the ENTSO-e system).[59]

(iv)   Assess the requirements for centralized SCADA systems that can be replicated as a backup in at least two locations.

(v)   Investigate cybersecurity issues and cloud storage alternatives for GMS technical data that provide a secure backup for existing national systems.

(vi)   Investigate alternative ways of achieving synchronous interconnections including the use of HVDC-VSC with synchronizing torque capability and VFT technologies as a means of stabilizing grid–to–grid operations prior to expanding the use of HVAC links.

---

[59]   ENTSO-e. 2021. *Continental Europe Synchronous Area Separation on 8 January 2021: Interim Report.* https://eepublicdownloads.azureedge.net/clean-documents/Publications/Position%20papers%20and%20reports/entso-e_CESysSep_interim_report_210225.pdf.

Appendixes

Appendix 1: GMS Energy Sector Assessment and Road Map

Milestone	Activities	Schedule
Study to identify the regulatory barriers to the development of power trade and implementation of Stage 2	Complete the study to identify the regulatory barriers to develop power trade and consider for adoption the measures and institutional arrangements to address regulatory barriers	2016
Study on a GMS Grid Code (operational procedures, performance standards, technical requirements, and regional planning)	Complete the study on a GMS grid code and consider for adoption the findings of the study, which include (i) GMS performance standards; (ii) coordination procedures between, system operators to schedule and control across border flows, management of deviations; (iii) metering and communications; (iv) process and analytical framework for process regional planning; and (v) sharing of power reserves and supporting during emergencies.	2017
Study on Stage 2 Transmission Regulations to allow third-party access in interconnections, with priority to contracts or PPAs, including Stage 2 power trade rules, and a Dispute Resolution Mechanism	Complete the study on Stage 2 transmission regulations and consider for adoption the findings of the study, which include (i) development of payment agreements or tariffs for third-party use, to compensate countries that host flows linked to third parties' trading; (ii) power trade rules for short-term cross-border trading; and (iii) power trade rules for settlement of deviations to scheduled power trade in grid-to-grid interconnections.	Mid-2019
Final review of Stage 2 readiness	Complete necessary legal, regulatory, and operation procedures to launch Stage 2	September 2019
Launch of Stage 2	Grid–to–grid power trading between any pair of GMS countries using transmission facilities of a third regional country	January 2020

GMS = Greater Mekong Subregion, PPA = power purchase agreement, TPA = third-party access.
Source: Asian Development Bank.

Appendix 2: ADB Technical Assistance and Knowledge Products on Power Trade in the GMS

Table A2.1: ADB-Funded Technical Assistance for Power Trade in the GMS

TA	Year	RPTCC WG/TA Consultant	Brief Description of Scope of TA Component	Independent Review of TA
TA5920-REG	2000		Regional Indicative Master Plan on Power Interconnection in the Greater Mekong Subregion	
TA6100-REG	2003 to 2005		Study for a Regional Power Trade Operating Agreement in GMS	
TA6304-REG	2006 to 2008		GMS Power Trade Coordination and Development	
TA6440-REG	2008 to 2010	RTE International, FRANCE; EDF-CIH, Hydro Engineering Centre; Nord Pool Consulting AS (NPC); Power Planning Associates UK, Franklin Paris, Legal Firm; Centre for Energy Environment Resources Development (CEERD)	Facilitating Regional Power Trading and Environmentally Sustainable Development of Electricity Infrastructure in the Greater Mekong Subregion in two components including Component 1 composed of five modules: (i) Regional Power Interconnection Master Plan (EDF-CIH) (ii) Methodology for Assessment of Benefits of Power Interconnection (NPC) (iii) Power Transmission Studies (RTE & PPA) (iv) GMS Regulatory Framework (RTE, NPC, PPA, Franklin) (v) Update of the Structure of the Existing Regional Database (RTE) Component 2 composed of two modules: mainly analysis/ studies/ evaluation and capacity building on: (i) Strategic Environmental Assessment (CEERD) (ii) Environmental Impact Assessment (CEERD)	Sida Review December 2011 (Juhani Antikainen, Rita Gebert, Ulf Møller) concluded TA was too broad and did not achieve its objectives to stimulate power trade. ADB review establishes RETA 8830.
RETA-7764	2011 to 2012		Ensuring Sustainability of GMS Regional Power Development To integrate environmental and social components into the ADB/GMS RETA No 6440 Master Plan. Training in use of Optgen software used to develop the power planning database under RETA6440 was also used under this TA	ADB TACR 439293 TCR 2015 outcomes not fully achieved due to lack of data for extensive power systems like the PRC.

Continued on the next page

Table A2.1 *continued*

TA 8356-MYA	2015	IES/MMIC	Energy Master Plan defines a long-term optimal fuel supply mix considering a country's primary resource endowments	Not reviewed
	2017	Bruce Hamilton, Adika	Updated National Power Expansion Plan for Myanmar; use of WASP or GT max software to review options for Myanmar generation expansion plan	
RETA 8830	2015 ongoing	Michael Caubet	1 RPCC established and operations commenced through continued guidance by the RPTCC 2 GMS performance standards and grid codes considered by WGPG 3; Guidelines for GMS regulatory framework proposed by WGRI	Dropped
TA 9003			Renewable Energy and Energy Efficiency Development in GMS	
TA 9426	2018	In progress	Preparing the Northern Cross-Border Power Trade and Distribution Project (TRTA LAO 51329-002)	
	2018		Lao PDR: Energy Sector Assessment, Strategy and Road Map	

Source: Asian Development Bank.

Table A2.2: Some ADB Knowledge Products Relevant to Power Trade in the GMS

Report Number	Title
SES:REG 2012–2018	Knowledge Products and Services: Building a Stronger Knowledge Institution or Special Evaluation Study
ISBN 978-92-9257-726-1	Energy Storage in Grids with a High Penetration of Variable Generation
ISBN 978-92-9261-471-3	Handbook on Battery Energy Storage Systems
ISBN 978-92-9254-986-2	Knowledge and Power: Lessons from ADB Energy Projects
TA 7764-REG	Ensuring Sustainability of Greater Mekong Subregion Regional Power Development; Part 1: Integrating Strategic Environmental Assessment into Power Planning
ISBN- 978-92-9262-038-7	Harmonizing Power Systems in the Greater Mekong Subregion: Regulatory and Pricing Measures to Facilitate Trade

Source: Asian Development Bank.

Appendix 3: Technical Assistance by ADB Development Partners

International Financing Institutions

Although the Asian Development Bank (ADB) is recognized as the lead international financial institution for the economic development of the Greater Mekong Subregion (GMS), important techno-economic studies were also undertaken by the World Bank and its partner agency the Energy Sector Management Assistance Program, other international institutions including the United Nations Economic and Social Commission for Asia and the Pacific (UNESCAP) and the International Energy Agency (IEA), along with bilateral agencies to support the GMS power trade development program. The most recent report by UNESCAP includes a strategy toward interconnecting regional grids such as the GMS with other regional groups within Asia and the Pacific.[1]

In 2016, the World Bank funded two country studies on GMS energy resource and transmission planning (i) on the energy resources in Myanmar; and (ii) building on international experience of viable smart grid solutions to support the development of Viet Nam's transmission network. The World Bank's most recent report (2019) covers opportunities to enhance power market integration in the GMS by investigating the feasibility of 10 different interconnection proposals.[2] These reports included a study of GMS national grid codes financed by German aid (GIZ) for the ASEAN Centre for Energy (ACE).[3] The Japan International Cooperation Agency (JICA) has also been providing significant technical support to the Lao PDR. Its most recent report describes a proposed Lao People's Democratic Republic (Lao PDR) national transmission plan including consideration of how the country might interconnect with its GMS neighbors.

The United States Agency for International Development (USAID) is currently recruiting consultants to make recommendations to determine the minimum Regional Grid Code (RGC) requirements in the Southeast Asia to facilitate bilateral and multilateral power trade and increase the ability of Asian power systems to accommodate higher levels of variable renewable energy (VRE).[4] A study financed by USAID has a comprehensive analysis of the interconnection issues between the

---

[1]   ESCAP. 2019. *Electricity Connectivity Roadmap for Asia and the Pacific.*

[2]   S. Thorncraft. 2019. *Greater Mekong Subregion Power Market Development: All Business Cases, Including the Integrated GMS Case.* Ricardo Energy & Environment: World Bank. https://documents1.worldbank.org/curated/en/541551554971088114/pdf/Greater-Mekong-Subregion-Power-Market-Development-All-Business-Cases-including-the-Integrated-GMS-Case.pdf.

[3]   ACE, GIZ, T. Ackermann, E. Troester, and P.-P. Schierhorn, eds. 2018. *Report on ASEAN Grid Code Comparison Review.* Jakarta. October. https://aseanenergy.org/report-on-asean-grid-code-comparison-review/.

[4]   R. R. Panda. 2018. Trans-Regional Energy Connectivity Between the ASEAN Power Grid and the South-Asia Power Grid: Prospects and Opportunities. Presentation at the ASEAN Power Grid Summit 2018. Vientiane, Lao PDR. 21–23 May. https://sari-energy.org/wp-content/uploads/2018/05/PRESEN1-2.pdf.

South Asian Association for Regional Cooperation (SAARC) and the Association of Southeast Asian Nations (ASEAN) regions that may be of relevance to planning extensions to GMS networks. In a recent development, JICA–USAID have signed an agreement to cooperate in assisting Thailand and Viet Nam by providing institutional support to develop a power trading regime in the GMS region.[5]

Other technical organizations have funded studies in the GMS, some of which have involved reviews of the applicability of ASEAN grid codes to accommodate new technologies. A reference list of the most recent reports produced by these organizations is given in Appendix 8. These include important contributions by the International Renewable Energy Association (IRENA), the National Renewable Energy Laboratory (NREL), Berkeley National Laboratory, and Energies Review as described below.

International Energy Agency Publications

Three 2019 IEA studies are relevant to GMS countries. The first examines several international jurisdictions that are similarly placed in an early stage of developing power trading.[6] The first IEA report notes that the role of regional institutions is critical and that it is possible to integrate power systems across borders without sacrificing local autonomy, though some balance between regional and local priorities is necessary to realize the full benefits of cross-border integration. The second report makes specific recommendations for the GMS to make progress based on the publication of the RGC.[7] The third IEA study provides a review of the extent of the People's Republic of China's regulatory and technical capability to integrate its local and regional resources.[8] IEA also has a Partner Country series to promote the integration of VRE. Its most recent study covers Thailand and suggests there are issues in the grid codes used by transmission system operator (TSO) and distribution system operators (DSOs) that could be improved to accommodate the increasing amount of wind and solar photovoltaic (PV) and seek consistency and alignment across all grid codes.[9] Although there is a broad set of detailed technical standards in an RGC, they do not sufficiently define the technical performance desired from VRE integration. It notes for example that the Energy Generating Authority of Thailand (EGAT), the TSO responsible for transmission system operation, has a different grid code that does not allow reverse power flow from the distribution network to the transmission network.

---

[5]   United States Embassy and Consulate in Vietnam. 2020. Mekong-U.S. Partnership Joint Ministerial Statement. 15 September. https://vn.usembassy.gov/mekong-u-s-partnership-joint-ministerial-statement/.

[6]   IEA. 2019. *Integrating Power Systems Across Borders*. June. https://www.iea.org/reports/integrating-power-systems-across-borders; A later IEA paper is also relevant to the GMS: IEA. 2019. *Establishing Multilateral Power Trade in ASEAN*. September. https://www.iea.org/reports/establishing-multilateral-power-trade-in-asean.

[7]   IEA. 2019. *Establishing Multilateral Power Trade in ASEAN*. September. https://www.iea.org/reports/establishing-multilateral-power-trade-in-asean.

[8]   IEA. 2019. China Power System Transformation. Assessing the benefits of optimised operations and advanced flexibility options.

[9]   IEA. 2018. *Partner Country Series--Thailand Renewable Grid Integration Assessment*. October. https://www.iea.org/reports/partner-country-series-thailand-renewable-grid-integration assessment.

Technical Institutions

The are several international institutions supporting and investigating regional power trading in ways working closely with member countries, particularly regarding the development of VRE and related projects that promote atmospheric carbon reductions. The main reports of interest are issued by IRENA, NREL, European Network of Transmission System Operators (ENTSO-e) along with several other associations listed in Appendix 7. Based on information on their respective web pages, the mandate of these three institutions covers the following:

International Renewable Energy Association. IRENA is an intergovernmental organization that supports countries in their transition to a sustainable energy future. It promotes the widespread adoption and sustainable use of all forms of renewable energy, including bioenergy, geothermal, hydropower, ocean, solar and wind energy, in the pursuit of sustainable development, energy access, energy security,[10] and low-carbon economic growth and prosperity. IRENA's document *Scaling Up Variable Renewable Power: The Role of Grid Codes* provides insights into further developments of the RGC.

National Renewable Energy Laboratory. The NREL focuses on creative answers to today's energy challenges. It advances the science and engineering of energy efficiency, sustainable transportation, and renewable power technologies and provides the knowledge needed to integrate and optimize energy systems. It is currently working on a program called Creating Consensus in Grid Modernization that includes the development of several new grid codes for cybersecurity, microgrid standards, energy blockchains, inverter-based resources, and new standards for grid data. Most recently, the NREL has funded a review of how to develop an electricity storage market in India that could well be a basis for a similar development in the GMS.[11]

European Network of Transmission System Operators. ENTSO-e and its TSO members are addressing key challenges relating to the development of market mechanisms from design elements to network code implementation and transparency publications. Given the considerable range of actions needed to create a harmonized market, the Secretariat's team has a broad remit including market integration and congestion management, enhancing regional cooperation, balancing and ancillary services markets, integration of renewable energy sources, market-related network codes, European transmission tariffs, inter-TSO compensation, a transparency platform for European electricity market information, and electronic data interchange.

---

[10]   IRENA. 2016. *Scaling Up Variable Renewable Power: The Role of Grid Codes*. May. https://www.irena.org/publications/2016/May/Scaling-up-Variable-Renewable-Power-The-Role-of-Grid-Codes.

[11]   I. Chernyakhovskiy et al. 2021. *Energy Storage in South Asia: Understanding the Role of Grid- Connected Energy Storage in South Asia's Power Sector Transformation*. NREL. https://www.nrel.gov/docs/fy21osti/79915.pdf.

Appendix 4: GMS Interconnection Planning Studies

Overview

Over the last 20 years, there have been several economic studies of energy resources in the Greater Mekong Subregion (GMS) to determine how they can be optimized by exploiting power trade opportunities within the region. The two most recent planning studies with these objectives have modeled the capital and operating costs of generation and transmission expansion plans for each country along with associated carbon emissions up to 2035 to determine the net present value (NPV) of economic benefits attributable to grid–to–grid interconnections.[1] A World Bank study (2019) based on economic assumptions prevailing in 2016, examined 18 proposed interconnections and evaluated business cases for 10 of the most promising transmission projects. The study also considered the practical issues of achieving regional synchronization to reap the benefits of fully integrated grid operations. The complementary Asian Development Bank (ADB) Manitoba Hydro International (MHI) Master Plan study (2021), better reflecting the post-coronavirus disease (COVID-19) situation, examined the economic sensitivities of regional strategies to facilitate synchronous power trading to determine a robust least-cost plan for developing grid-to-grid interconnections. A summary of both reports is provided here.

Greater Mekong Subregion Power Market: All Business Cases (Including the Integrated GMS Case)

This World Bank study considered economic, technical, commercial, and environmental issues using the least-cost modeling PROPHET software to establish NPV costs and benefits for each of the 10 promising projects.[2] The study also compared the results with an alternative Integrated GMS Case where project staging, costing about $4.5 billion by 2035, are prioritized. The proposed grid–to–grid interconnections were assumed to be capable of transmitting 300–2,400 megawatts (MW) growing in increments up to 2035. When each project was modeled separately, the resultant power trading indicated the positive NPVs ranged from $36 million to $1.6 billion. The integrated case is based on the model developing an optimal plan of expansion until all the proposed interconnections were in place. Its cost was 30% lower

---

[1] NPV of the economic generation and grid benefits of interconnections less the up-front costs of the interconnecting lines and terminals and associated grid strengthening up to 2035. The NPV values computed in each study are significantly different: World Bank used a 10% discount rate applied from 2016 for 20 years. ADB used a lower rate (8%) applied over a shorter period from 2022 for 12 years. . Thorncraft. 2019. *Greater Mekong Subregion Power Market Development: All Business Cases, Including the Integrated GMS Case.* Ricardo Energy & Environment: World Bank. https://documents1.worldbank.org/curated/en/541551554971088114/pdf/Greater-Mekong-Subregion-Power-Market-Development-All-Business-Cases-including-the-Integrated-GMS-Case.pdf; and ADB. 2020. *Regional Power Master Plan: Harmonizing the Greater Mekong Sub-Region (GMS) Power System to Facilitate Regional Power Trade.* Consultant's report. Manila. https://www.adb.org/projects/documents/reg-47129-001-tacr.

[2] S. Thorncraft. 2019. *Greater Mekong Subregion Power Market Development: All Business Cases, Including the Integrated GMS Case.* Ricardo Energy & Environment: World Bank. https://documents1.worldbank.org/curated/en/541551554971088114/pdf/Greater-Mekong-Subregion-Power-Market-Development-All-Business-Cases-including-the-Integrated-GMS-Case.pdf.

than the sum of the individual project investment costs, supporting the notion that prioritization of the interconnections will be key in maximizing benefit across the GMS.

The study developed a comprehensive plan for regional interconnection and synchronization in four stages, each associated with an NPV for generation and transmission costs that exceeded the base case with each GMS country grid independently continuing to carry out its current generation and transmission plans. The stages of synchronism and demand growth included the following:

Stage 1 (2022–2024). Synchronization of a portion of the power systems in the southern part of the Lao People's Democratic Republic (Lao PDR) and central Viet Nam, building upon the existing synchronous connection between southern Cambodia and Viet Nam.

Stage 2 (2025–2027). Formation of four synchronous interconnections: (i) Viet Nam and parts of Cambodia and the Lao PDR, (ii) the Lao PDR with its medium voltage connections to Thailand, (iii) the Lao PDR and Cambodia, and (iv) the Lao PDR and Thailand.

Stage 3 (2028–2030). Formation of eastern and western GMS synchronized.

Stage 4 (2030–2035). Integration of GMS western and eastern synchronous interconnections to have a fully synchronized GMS power system.

The results and recommendations of the World Bank study are presented in charts showing the ranges of NPVs for each business and the integrated cases showing transmission investment costs, generation, and emission cost differences. The study included a recommended interconnection expansion plan based on a strategy for regional synchronization. As shown in Table A4.1, the ratio of the NPV of the benefits to the costs of interconnections ranges between 3 and 4.

Table A4.1 A Consolidated Summary of NPV Benefits by the World Bank

Summary of World Bank Report Annex A Table 26		Net Present Value (10% rate) of Benefits Across All Business Cases (million, real 2016)					
Business Case	Size (MW)	Generation Capex	Generation Operations	Grid Reinforce	Intercon-nections	Sum Total Net Benefits	Ratio NB/ Inter Costs
Total Eight ASEAN Projects	Stage I	2,410	3,714	256	(1,012)	4,435	4.38
	Stage II	3,816	8,966	174	(2,273)	9,750	4.29
	Stage III	2,360	15,842	54	(3,667)	13,656	3.72
Integrated (China #$50/ MWh)	Stage IV	(640)	3,927	73	(899)	2,461	2.74
Total Development Plan	by 2030	1,720	19769	127	(4,566)	16,117	3.53

ASEAN = Association of Southeast Asian Nations, Capex = capacity expansion, MWh = megawatt hour, NB = net benefit, MW = megawatt.

Source: S. Thorncraft. 2019. Greater Mekong Subregion Power Market Development: All Business Cases, Including the Integrated GMS Case. Ricardo Energy & Environment: World Bank; and ADB. 2020. Regional Power Master Plan: Harmonizing the Greater Mekong Sub-Region (GMS) Power System to Facilitate Regional Power Trade. Consultant's report. Manila.

The PROPHET model also computed the regional benefits with emissions falling approximately 7% over the period from 2020 to 2035 associated with a reduction in coal-fired generation across the GMS. It showed that by 2035, the integrated case results in 670 million tons of carbon dioxide equivalent emissions, down from 730 million tons in the base case. The study demonstrated that the highest benefits (with ratios of benefits/connection cost ratios between 6 and 8) will come from greater integration of the Lao PDR with its neighboring countries where the Lao PDR could play a role in terms of providing additional power supplies to Myanmar and Viet Nam with immediate short-term cost reductions, and over the longer-term for Thailand. The recommended GMS-wide synchronization plan, however, would need to be reviewed considering the current plans by the new transmission company in the Lao PDR Électricité du Laos Transmission Company Limited (EDL-T) to integrate its internal networks and synchronously interconnect them with the People's Republic of China (PRC).

Regional Power Master Plan

The MHI Master Plan was developed by modeling the GMS transmission networks, along with all existing, under construction, and potential generating plants.[3] It used the OptGen or SDDP model along with Power Systems Simulation Software in common use in ASEAN region software to analyze the impact of power trading on the national grids to determine if additional grid transmission reinforcements were necessary. The power system model for the PRC (Yunnan and Guangxi provinces) is represented by a small number of nodes, and it was assumed that the PRC would not be able to export any surplus for several years. Accordingly, power trade with the PRC was modeled by connecting from the designated nodes to the Lao PDR, Myanmar, and Viet Nam with lumped generation or demand to model the cross-border power trade.

The model was initialized with 18 known transmission candidate projects, similar to those used in the World Bank study, with transmission lines ranging between 150–1,300 kilometers (km) in length, capable of transferring 300–3,000 MW in stages up to 2030. As the modeling optimization progressed, several other interconnections were added to represent an aggressive cross-border transmission development scenario. In total, 36 regional generation planning scenarios were developed for the period from 2022 to 2035 considering high-, medium-, and low-demand growth conditions. In addition, 12 "sensitivity scenarios" were identified to cover unforeseen and extreme future situations including COVID-19 where regional demand was expected to decrease by 8%. The other scenarios included low future hydro availability (by 20%), reduced carbon scenarios, and a scenario with no import restrictions currently imposed by Thailand and Viet Nam. A significant portion of the report includes lists of generation and transmission data intended for use in evaluating future options. The Regional Power Trade Coordinating Committee (RPTCC) could do this in the future by updating the assumptions regarding high

[3]   ADB. 2020. *Harmonizing Power Systems in the Greater Mekong Subregion: Regulatory and Pricing Measures To Facilitate Trade*. Manila. Appendix 3. http://dx.doi.org/10.22617/. MHI also submitted a brief report titled Additional Analysis dated 8 July 2021 providing ADB with more definitive information relating to their recommended optimal interconnection investment program.

medium and low demand scenarios, economic factors (high fossil fuel, low variable renewable energy [VRE], and low gas prices), and technological factors (aggressive power trading, nuclear energy, and battery storage).

Taken together, the data and assumptions were applied in the model to develop national generation and transmission development scenarios (only for the five ASEAN member countries of the GMS), along with identified cross-border transmission connections with the nominated PRC nodes. The study provided outcomes for the scenarios of demand growth in terms of the assumed economic and technological factors with the transmission optimization process ensuring the transmission reliability in system-intact (N-0) operation. The medium load base scenario was analyzed in more detail to verify the reliability of a cross-border transmission network for an N-1 transmission outage conditions. The objective of this analysis was to identify potential transmission congestion situations and identify high-level mitigation options. The solutions proposed here should be further studied and refined in future feasibility studies.

The report shows that the development of cross-border interconnections under the medium demand scenario would result in an 8-12-gigawatt (GW) reduction of new thermal generation development in the GMS up to 2035. It showed that the development of large-scale coal and gas plants in Thailand and Viet Nam could be avoided or delayed with the use of cross-border interconnections. In contrast, hydro generation in Myanmar and the Lao PDR increased after enabling cross border interconnection optimization. Table A4.2 provides a comparison of the sensitivity scenarios.

Table A4.1 shows a net benefit for the medium demand base case of $20 billion compared with the reference scenario without grid-to-grid interconnection. The four-to-six ratio of net benefit–cost of interconnections compares favorably with the three-to-six ratio estimated in the World Bank study. The results are robust under the sensitivities considered. Both the reduced gas price scenario (S1) and the aggressive interconnection development scenario (S8) show higher cost savings compared to the base scenario (BM). A scenario with both reduced VRE costs and aggressive interconnection development (S10) does not show any significant additional savings. High fossil fuel price scenario (S7) results in an increased cost of $13 billion due to the increased operating cost. Such increments due to fuel cost can be offset by facilitating increased cross-border power trade as shown in the scenario "high fossil fuel price and high cross-border power trade" (S9). The MHI's Additional Analysis provides specific details of the base case with the building of 14 interconnections ranging in capacity from 300 MW to 4.2 GW with an aggregate transfer capacity of 12 GW, altogether costing about $3.3 billion. Under a proposed Accelerated Scenario, 24 connections would be built ranging from 4.5 GW to 7.9 GW with a total aggregate capacity of 33.3 GW costing $4.5 billion.

The study provided an indicative plan for synchronization pointing out the necessity for comprehensive technical studies to ensure this could be done safely without undue disturbance to existing grids. It considered the challenges of achieving synchronous high voltage alternating current (HVAC) interconnections, recognizing there are significant differences in the dynamic performance of national grid operations that would require compliance with the Regional Grid Code (RGC) before they were interconnected. It noted, however, that asynchronous high voltage direct current (HVDC) systems have a number of advantages well-suited for cross-border interconnections such as power transfer flexibility, fast control, and blocking capability. If the necessary generation dynamic performance improvements

Table A4.2 GMS MHI Master Plan Cost–Benefit Summary

MHI Scenario ID	Scenario Description: RM: S8 Medium Demand Forecast with Associated Sensitivity Scenario: RH: BH High Forecast, RL, BL Low Forecast	Generation Capital Expansion 2022-2035		Net Present Value US$b at 2022 (8% discount rate)					Benefit Ratio
				Generation		Trans			
		GW	US$b	Capex	O&M	Capex	Total Gen and Trans	Net Benefit to Ref	Benefits Trans-mission Costs
RM	Reference scenario – Medium Load Forecast	74	127	50	212	2	52		
BM	Base scenario (most likely development scenario)	73	120	51	190	4	55	19	4.75
S1	Reduced gas price	69	111	46	173	4	50	41	10.25
S2	Reduced VRE costs	81	119	55	181	4	59	25	6.00
S3	High fossil fuel prices	80	124	62	207	5	67	(10)	(2.00)
S4	Battery storage systems (BSS) for solar	69	120	51	189	4	55	20	5.00
S5	Reduced VRE costs and BSS for Solar	83	124	55	183	4	59	22	5.50
S6	Nuclear generation option	86	158	48	190	4	52	22	5.50
S7	High fossil fuel price with nuclear generation option	86	158	72	205	5	77	(18)	(3.60)
S8	Aggressive cross-border power trade	77	119	51	177	5	56	31	6.20
RH	Reference scenario – High Forecast	108	174	80	262	2	82		
BH	Base scenario (most likely development scenario)	110	176	80	216	5	85	43	8.60
RL	Reference scenario – Low Forecast	41	66	28	187	0	28		
BL	Base scenario (most likely development scenario)	47	75	27	168	4	31	16	4.00

Capex = ?, GMS = Greater Mekong Subregion, GW = gigawatt, O&M = ?, US = United States, VRE = variable renewable energy (typically wind and solar power).

Source: ADB. 2020. *Regional Power Master Plan: Harmonizing the Greater Mekong Sub-Region (GMS) Power System to Facilitate Regional Power Trade.* Consultant's report. Manila.

are not completed and there is delayed commissioning of the HVAC links, the estimated loss in NPV benefits is $11 billion. Accordingly, the study recommended that HVDC cross-border interconnections be considered further to accrue the benefits of cross-border power trade as early as possible.

Summary and Conclusions

The extensive modeling in both these studies confirmed that GMS grid-to-grid power trading will benefit all countries within the region. An investment of about $3 billion to $4 billion in transmission interconnections would provide 4–6 times the cost in avoided planned generation investments, deferring the need for national transmission upgrades, and avoiding higher power generation costs. Other identified benefits include being able to deploy higher levels of renewable energy in the region, better use of hydro resources because of diversification in hydrological conditions, and diversification in demand profiles of the interconnected countries. These benefits arise from only a limited number of interconnections that were studied, specifically relating to Myanmar to the PRC and the Lao PDR, and the Lao PDR to Viet Nam.

Both studies note that it is essential for the RPTCC to develop a conceptual road map to facilitate the progression of the GMS toward a more tightly integrated synchronized power system, preferably before 2030. However, it is feasible to commence power trading as soon as possible, using HVDC asynchronous interconnections initially to facilitate sharing reserve capacity. The GMS needs to prioritize the staged planning of the proposed interconnection as soon as possible. This will enable the greater use of HVAC connections in transmission along with lower voltage connections in distribution networks, particularly where the latter supply remote villages across borders. To mitigate the concerns of country variable renewable energy (typically wind and solar power) in implementing the synchronization program, it will be necessary to obtain further technical assistance support to carry out extensive power system load flow, fault, and stability studies using actual system data to identify the critical components that may need to be brought into compliance with the RGC.

Appendix 5: Worldwide Regional Power Trading Regimes

International grid interconnections can be as modest as the one-way transfer of a small amount of electricity from one country to another, or as ambitious as the full integration of the power systems and markets of all the countries in a region. Currently, there are a number of grid-to-grid interconnections in regions around the world either operating or in the planning phase, including the following:[1]

(i)   **European Union and the United Kingdom.** The region includes Western and Central Europe, (UCTE-e); Scandinavia (NORD POOL); the Southeastern Europe regional market; and the Baltic electricity market.

(ii)   **North America.** The United States (US) and Canada have extensive regional electricity and gas trade markets, not all of which are interconnected. The regional synchronous interconnections cover both Canada and the US, with some interconnections to Mexico.

(iii) **Russian Federation.** Spanning eight time zones, the Russian Federation has operated a centralized system for many years within its IPS or UPS synchronization zone characterized by a high degree of interdependence and cooperation. Strictly speaking, while not regarded as a competitive, open access power market, it has all the technical aspects of a well-coordinated power system operation.

(iv)   **People's Republic of China.** The world's largest integrated high voltage direct current (HVDC)/high voltage alternating current (HVAC) power system with provincial authorities trading among themselves and with countries and regions on their borders.

(v)   **Gulf Cooperation Council Interconnection Authority.** The Gulf Cooperation Council Interconnection Authority (GCCIA) uses a backbone 400-kilovolt grid to interconnect the 60-hertz Kingdom of Saudi Arabia to the six bordering 50-hertz Gulf countries by HVDC. The Pan Arab Grid Code was published in February 2020 and is expected to be the basis for expansion of trade with the European Network of Transmission System Operators (ENTSO-e) market.

(vi)   **South American Electricity Markets.** Southern Cone countries and the Andean Community of Nations.

(vii) **Africa.** The Southern African Power Pool and Western African Power Pool had some interconnections and planned interconnections in the Nile River Basin, West Africa, and East Africa.

(viii) **India.** India has recently achieved a nationwide interconnection and is pursuing further integration with South Asian Association for Regional Cooperation (SAARC) countries. This region has aspirations to trade with

---

[1]   This list is an updated version of the information provided in the UN Economic and Social Affairs Division for Sustainable Development published in 2005 titled *Multi-Dimensional Issues in International Electric Power Grid Interconnections.* The 200-page document covers many aspects of grid-to-grid interconnections including technical, economic, financial, legal, political, social, environmental, and energy security.

the Greater Mekong Subregion (GMS) through interconnections between Bangladesh and Myanmar.

(ix)   **Commonwealth of Independent States.** The countries of the former Soviet Union are planning to increase reintegration and trade with each other and neighboring markets. The region is currently building an HVDC interconnection with SAARC from Tajikistan to Pakistan.

(x)   **Central America.** Central American Electrical Interconnection System.

(xi)   **Southeast Asia.** The GMS initiative for integrated electricity markets and the Association of Southeast Asian Nations are developing cross-border interconnections and looking into opportunities to enhance the security of supply and renewables development by working together.

Appendix 6: Overview of GMS Transmission Network, Power Sector Status, and Development Plans

Country or Region	Key Features of Transmission Network and Interconnectivity with GMS	Main Features of Current Power Development Plans	Key Challenges (includes impact of COVID-19)
PRC (CSG) MD 47 GW 53% coal and gas 40% HPP	• 500/800 kV AC and DC OHTL • 8 lines with GMS: 2@500 kV Myanmar, 1@115 kV Lao PDR; 3@22 kV; 3@110 kV Viet Nam • Plans to export to Bangladesh and Thailand via Myanmar • HVDC interconnection with Northern Grid	• Reduce carbon intensity • Reduce share of coal • Promote private sector with feed-in tariffs for developments in low-carbon tech, including HPP, nuclear, and VRE • 2030 target to have 20% renewables • Expanding electricity trade with ASEAN	• Mobilizing generation in the west to load centers in the east • Slowdown of economic and electricity demand growth • growing surplus of generation
Cambodia MD 1.75 GW 31% coal 55% HPP 13% fuel	115/220 kV OHTL interconnect grids of the national system • Grid-to-grid with the Lao PDR, Thailand and Viet Nam via 220/230 kV, 115 kV and • 30 import connections at 220 kV • plans to adapt GMS RGC 12/9/20	• Develop new generation capacity from coal and HPP and gas in the longer term • High growth (22%) plan to increase grid electricity access to 70% by 2030 • Reduce electricity prices • Enhancements to the transmission system to develop a stronger gird	• Low electrification rate • High electricity prices • Reducing reliance on power imports in the short term
Lao PDR MD 0.9 GW 30% coal 70% HPP	• Four connected regions: north, central 1 and 2 and south • 115 kV OHTL connecting between hydropower plants and loads • cross-border electricity trade with dedicated hydro and lignite power projects in Thailand and Viet Nam • Several grid-to-grid connections at low and medium voltage levels for importing from the PRC and Thailand as well as exporting to Cambodia	• Expand generation capacity to deliver reliable, sustainable, and affordable electricity • Increase electrification to 90% by 2020 • Promote power exports to earn more revenue for poverty reduction • Promote deployment of small hydro, solar, wind, and biofuels • Improve OHTL in the northern, central, and southern areas and links to Thailand and Viet Nam under JICA PDMP	• High dependence on hydropower • Unable to exploit benefits from geographical diversification of hydro generation in the north and south given the absence of north-south transmission interconnections • Considering 500 kV/HVDC backbone from the PRC to Cambodia
Myanmar MD 1. 64 GW 61% HPP 30% gas	• 132 kV and 230 kV transmission backbone connecting two major load centers in the central north and south	• A target to achieve 100% electrification by 2030 • MOEP's plan suggests long-term capacity mix dominated by hydro, coal, gas, and VRE.	• Low electrification rate • High-demand growth is expected to continue to 2030 • Unreliable supply resulting in power supply shortages

Continued on the next page

Appendix 6 *continued*

	• HVAC and HVDC cross-border OHTL to export electricity from dedicated HPP to Yunnan in the PRC • Recent tenders for 1000 MW PV up to 30 MW per IPP	• PDP evolving with the optimal generation mix being strongly debated. • Risk of shortages, long lead times with gas availability being limited and minimal generation investment in the near term.	
Thailand MD 30 GW 18% coal 60% gas 13% HPP	• 230 kV and 500 kV transmission backbone for transferring power from generation sources in the north, west and northeast to load centers in the central area • HVAC cross-border transmission lines for importing power from dedicated power projects in the Lao PDR • 115 kV and 22 kV grid-to-grid connections for exchanging power with the Lao PDR and exporting to Cambodia	• Enhance security of supply by diversifying the fuel mix and reducing the share of gas-fired electricity generation • Promote renewable energy and energy efficiency to reduce carbon intensity • Latest PDP (2015) suggests long-term capacity mix consisting of 30%–40% gas, 20% renewables, 20%–25% coal, 15%–20% hydro • Plans to develop RE forecasting center, technical issues with VRE at distribution level	• High dependence on natural gas for electricity generation, high fuel costs compared to the rest of the GMS • High reserve margin over the next 15 years due to lower-than-expected demand growth and excessive investment in generation capacity
Viet Nam MD 35 GW 37% coal 45% HPP 16% gas	• High voltage transmission backbone connecting north, central, and south of Viet Nam • Cross-border power trade exists with Cambodia, the Lao PDR, and the PRC • 220 kV grid–to–grid isolated connections for importing power from the PRC and exporting power to Cambodia • HVAC cross-border transmission lines for importing power from dedicated hydro projects in the Lao PDR	• Reforming the power sector to create competition and support IPPs • Latest PDP (2016) suggests long-term capacity mix consisting of 43% coal, 17% hydropower, 15% gas, 21% renewables and the rest from nuclear and imports • Revised stance on nuclear now not an option in long-term planning • Developments in the energy sector revolve around domestic and imported coal and off-share gas reserves • New MP prioritizing VRE	• High electricity demand growth, pressure on tariffs • Increasing reliance of imported coal and gas • Long lead times in developing offshore gas reserves • Technical challenges in implementing high VRE generation share into the system • Absorption of high levels of PV

AC = alternating current, ASEAN = Association of Southeast Asian Nations, COVID-19 = corona virus pandemic, DC = direct current, GMS = Greater Mekong Subregion, GW = gigawatts, HPP = hydropower plant, HVAC = high voltage alternating current, IPP = independent power producer, JICA = Japan International Cooperation Agency, kV = kilovolt, Lao PDR = Lao People Democratic Republic, MOEP = ministry of electric power, OHTL = overhead transmission line, PDMP = power master development plan, PDP = power development plan, PRC = Peoples Republic of China, RE = renewable energy, RGC = regional grid code, VRE = variable renewable energy.

Source: S. Thorncraft. 2019. *Greater Mekong Subregion Power Market Development: All Business Cases, Including the Integrated GMS Case*. Ricardo Energy & Environment: World Bank. https://documents1.worldbank.org/curated/en/541551554971088114/pdf/Greater-Mekong-Subregion-Power-Market-Development-All-Business-Cases-including-the-Integrated-GMS-Case.pdf. The table is adapted from Table 10 on page 42 of that report with updated information from the Regional Power Trade Coordination Committee country report.

Appendix 7: Main Tasks Proposed for Working Groups under Road Map Phase II

Achievements	Steps 1–4 Adapted and Updated from RPTCC 24 June 2018
1 Common Performance Standards	Initial document of GMS performance standards, 2010 • Review, Update 2012 to 2016 • Agreed to adopt as a reference document, June 2016
2 Transmission Regulations	Four Transmission Policies (drafts in 2014) Adopted as reference documents December 2017 • Policy on Scheduling and Accounting • Policy on Co-ordinated Operational Planning • Policy on Communication Infrastructure • Policy on Data Exchanges, Glossary and Definitions Summary Report
3 Metering	Standard Regional Metering Arrangements and Communications November 2017 • Comments provided by all countries, June 2018
4 Regional Grid Code	RGC accepted by RPTCC March 2019, published on website January 2021 • Master Plan feasibility studies of priority interconnections, Draft October 2020 Completed June 2021 • Expanding the WGPG to include the above tasks and progressing to next stages of planning and operationalizing the grid code.
Implementation of RGC	Steps 5–10 Adapted and updated from recommendation RPTCC 26 & 27 November 2019
5 Establish an implementation roadmap for the regional GMS Grid Codes including enforcement measures, review mechanism and structure, gap assessment for national regional compliance, and identification of mitigation measures.	Enforcement of the GMS Grid Code • Draft the Intergovernmental MOU to adopt the GMS Grid Code. • Establish governance structure for management of the GMS Grid Code, including the Grid Code Secretariat and the RGC Review Panel with effective RPCC setup. **Assess National Technical Framework for Compliance with the GMS Performance Standards, and Harmonization of Operational Practices.** • Technical assessment of generation, transmission, and distribution facilities to respond to the GMS performance standards and technical requirements of the GMS Grid Code. • Propose technical remedial measures with estimated costs and implementation plan. **Operational processes, procedures, and rules for compliance with the GMS Grid Code** • Assess existing national operational processes, procedures, and rules for compliance with the GMS Grid Code required practices. • Develop new required processes, procedures, and rules to follow the required practices prescribed by the GMS Grid Code for national purposes (security analysis, outage scheduling, national grid model, LFCR methodology and procedures, and restoration plans, balancing mechanism, etc.). • Propose technical remedial measures with estimated costs and implementation plan. • Develop a mechanism for compliance monitoring of the GMS countries at regional level.
6 Establish an organization for operationalization of GMS synchronous areas including LFC organization and structure, operational agreements, methodologies, processes and procedures.	LFC Organization and Structure • Assess LFCR capacities in each of the GMS countries. • Set up LFC organization and structure for the GMS Region in collaboration with the TSOs of GMS countries. Synchronous Area Operational Agreement • Assess the LFCR practices and needs in each of the GMS countries. • Establish proposed Synchronous Area Operation Agreement in collaboration with TSOs in GMS Countries.

Continued on the next page

Appendix 7 *continued*

	Multi-Parties Operational Agreements (LFC Block Operational Agreements, LFC Area Operational Agreements, Monitoring Area Operational Agreements, Cross-Border FRR Activation Agreement, Cross-Border RR Activation Agreement, Sharing Agreement, Exchange Agreement, etc.) • Assess the needs in terms of Multi-Parties Operational Agreements to be established. • Establish proposed Multi-Parties Operational Agreements in collaboration with TSOs of the concerned member countries. **Establishment of common methodologies for coordinating operational security analysis in each synchronous areas** • Establish seasonal peak generation adequacy outlooks. • Assess the relevance of power generating units, demand facilities, and grid elements for the outage coordination process, transmission capacity calculation, and definition of the low frequency demand disconnection scheme. • Assess the current practices to take into consideration the specificities of each GMS member country. • Establish proposed common methodologies in collaboration with the TSOs of the concerned member Countries. **Establishment of Coordination Procedures for: Common Operational Security Analysis; Regional Outage Coordination for Recovery and Restoration.** • Assess current practices to take into consideration the specificities of each GMS member country. • Establish proposed coordination procedures in collaboration with the TSOs of the concerned member countries.
7 Establish the design of regional electrical ITC system allowing a safe and secure exchange of data and information among GMS market operators and the RPCC.	**Preliminary Design of the Regional ITC System (Electronic Highway)** • Assess ITC systems in use in the GMS countries (protocol, capacities, speed, etc.). • Assess future needs in terms of capacity flows of data, information exchange, required speed, security of exchange to preserve the confidentiality of the information, recommended protocol, and etcetera. • Set up Preliminary Design of the Regional ITC System for Data & Information Exchange in collaboration with the TSOs of the member countries.
8 Establish an ad hoc metering organization and architecture allowing for an appropriate settlement of power exchange in the GMS, including meter equipment on tie-lines and a centralized data processing system.	**Preliminary Design of the Metering Organization and Architecture** • Assess the existing metering organization and architecture in use in the GMS countries. • Assess future needs in terms of power and energy-metered data on the interconnection tie-lines, meter specifications, centralized data processing system to be developed, necessary backup and remote reading systems, security of data exchange to preserve the confidentiality of the information protocol, and etcetera. • Set up the preliminary design of the metering organization and architecture to be developed in the GMS Region, taking into consideration the central role played by the RPCC, in collaboration with the TSOs of the member countries.
9 GMS regional master plan	• Review and adoption of strategic planning criteria • Regional simplified network model and establishment of a regional database for periodic review
10 New projects	• Assess new interconnection projects and manage the portfolio.

GMS = Greater Mekong Subregion, LFC = load frequency control, MOU = memorandum of understanding, RGC = regional grid code, RPCC = Regional Power Coordination Center, RPTCC = Regional Power Trade Coordination Committee, TSO = transmission system operator, WGPG = Working Group on Performance Standards and Grid Code.

Source: ADB. 2020. Summary of Recommended Actions Described in the RGC Consultant's Progress Report V8: Section 4 WGPO Tasks 5-10.

Appendix 8: List of Principal References

Date	Title	Reference: Agency/Consultant/ Website	Summary of Contents
Current	ADB website	https://www.adb.org/publications	Knowledge Products, GMS, and TA information.
Current	RPTCC Website	https://www.greatermekong.org/rptcc	Contains summaries of RPTCC meetings, MOUs, plus country submissions and presentations.
Current	ENTSO-e Website	https://www.entsoe.eu/network_codes/	Starting page for details of ENTSO-e Grid Codes.
Nov 2020	XFLEX HYDRO	https://www.hydropower.org/ Institute for Systems and Computer Engineering, Technology and Science (INESC TEC)	The European Union funded XFLEX HYDRO project explores the emerging opportunities for hydropower plants to provide new short-term flexibility and system support services, known as ancillary services, to the European grid.
Feb 2020	The Study on Power Network System Master Plan in the Lao People's Democratic Republic	Final Report JICA: by TEPCO power Grid/Nippon Koei/TEPSCO https://openjicareport.jica.go.jp/pdf/12328027_01.pdf	Publication of a proposed national transmission plan including consideration of how the Lao PDR will connect with its GMS neighbors. The report includes a special section (Chapter 7) giving details of a contrasting proposal by the PRC proposing to build HVDC and 500 kV through the Lao PDR by 2030.
Feb 2020	HPSGMR: Regulatory and Pricing Measures to Facilitate Trade	ADB knowledge product: TA 8830-REG	This addresses regulatory and pricing measures relating to (i) opening access to competitive generators, (ii) a method for wheeling charges, (iii) rules for short-term bilateral trading and, (iv) a balancing mechanism.
2020	Roadmap for multisectoral planning support	https://www.entsoe.eu/news/2020/07/16/toward-a-system-of-systems-entso-e-releases-roadmap-for-coordinated-multi-sectorial-planning/	ENTSO-e will improve the consideration of smart sector integration in the infrastructure planning process. Smart sector integration will enhance flexibility across energy sectors and allows for development toward a more energy- and cost-efficient energy system.
June 2019	Integrating Power Systems across Borders	IEA: https://www.iea.org/reports/integrating-power-systems-across-borders	Expanding power systems across borders allows developers and market participants to take advantage of economies of scale on both supply and demand sides thereby enabling greater resources and access to cheaper supply sources.

Continued on the next page

Appendix 8 *continued*

2019	Integration of Large-Scale PV-Based Generation into Power Systems: A Survey	Energies Review: https://www.mdpi.com/1996-1073/12/8/1425	General overview of current research on analysis and control of the power grid with grid-scale PV-based power generation and consequences of grid-scale integration of PV generation units into power systems.
2019	PRC Power System Transformation	IEA: https://webstore.iea.org/download/direct/2440	"Assessing the Benefits of Optimized Operations and Advanced Flexibility Options." Comprehensive review of the extent of the PRC's regulatory and technical capability to integrate its local and regional resources. Has many references to ongoing studies of regional grids.
2019	Solutions for a Renewable Powered Future	IRENA: https://www.irena.org/innovation/Solutions-for-a-Renewable-Powered-Future	This includes easy-to-read chapters on advanced weather forecasting, flexible generation to accommodate variability, Interconnections and regional markets as flexibility providers, matching renewable energy generation, and demand over large distances with super grids.
2019	GMS Power Market Development	World Bank – IES/NordPool, Ricardo Energy & Environment. Author Stuart Thorncroft	"Selected Business Cases for Power Trade" including the Integrated GMS Case on enhanced power market integration in the Greater Mekong Subregion.
2019	HPSGMR: Regional Master Plan	MHI interim report – Presented at RPTCC 19	Preliminary results of a technical study on the optimization of regional transmission and generation in the GMS. Final report due December 2020.
2019	International Grid Code Comparison Listing	DNV GL (Germany) GC Compliance Listing: http://www.dnvgl.com/GridCodeListing.pdf	Provides web page links to published NGCs and RGCs around the world.
2018	Challenges and Opportunities of LFC in Conventional, Modern and Smart Power Systems	Energies 2018: https://www.mdpi.com/1996-1073/11/10/2497	"A Comprehensive Review of the State of the Art of LFC" including HVDC as a method of synchronizing adjacent power systems.
2018	Review of International Grid Codes	Ciaran Roberts; Energy Analysis and Environmental Impacts Division Lawrence Berkeley National Laboratory	Summary of grid code practices in the Brazil, Canada, Finland, Italy, Ireland, New Zealand, Singapore, Spain, Switzerland, the UK, the US, and the summary covers grid usage, system services, grid expansion, and general grid operation.
2018	Report on ASEAN Grid Code Comparison Review	ASEAN Centre for Energy (ACE) & (GIZ) GmbH	Grid code recommendations provide basic criteria for system planning, connection, and operational rules and responsibilities to be followed by generating stations, transmission utilities, and distribution utilities.

Continued on the next page

Appendix 8 *continued*

2018	Thailand Renewable Grid Integration Assessment	IEA https://webstore.iea.org/partner-country-series-thailand-renewable-grid-integration-assessment	An analysis of existing VRE penetration in Thailand, the technical potential and economic impact of distributed solar PV on stakeholders, and the power sector planning process and system costs.
2016	Compare Grid Codes in the PRC and US	NREL: https://www.nrel.gov/docs/fy16osti/64225.pdf	"Comparison of Standards and Technical Requirements of Grid-Connected Wind Power Plants in PRC and the United States."
2016	Scaling Up Variable Renewable Power: The Role of Grid Codes	International Renewable Energy Agency – IRENA: https://www.irena.org/Publications	Provides guidance to regulators, policymakers, system operators and other stakeholders on how grid connection codes should be developed and implemented, including case studies.
2016	GMS: Energy Sector Assessment and Road Map	ADB GMS website	Highlights sector performance, priority development constraints, government plans and strategies, past ADB support and experience, other development partner support, and future ADB support strategy.
2016	Smart Grid to Enhance Power Transmission in Viet Nam	World Bank/ Energy Sector Management Assistance Program: CESI International https://openknowledge.worldbank.org/handle/10986/24027	Building on international experiences, the report identifies viable smart grid solutions for Viet Nam's transmission network. Makes recommendations for the use of HVDC to reinforce Viet Nam's existing 500 kV.
2013	GMS Energy Strategy Report:	ADB/CEERD	Assessment of the GMS Energy Sector Development Progress, Prospects and Regional Investment priorities.
2012	Summary of TA6440 Outputs ASEAN Officers of Electricity Regulatory Commissions	ADB/CEERD. Power Pont Report by Prof T Lefevre. 5th Capacity Building Program for	Facilitating Regional Power Trading and Environmentally Sustainable Development of Electricity Infrastructure in the Greater Mekong Subregion The Case of ADB/GMS RETA 6440 GMS Master Power Plan.
2011	Review of the Greater Mekong Sub-Region Regional Power Trade	Sida Review -Final Report Juhani Antikainen; Rita Gebert; Ulf Møller https://www.publikationer.sida.se/	Assessment of the outputs and outcomes of RETA 6440 concluding that the program lacked focus. The lack of transmission infrastructure in smaller countries, and the absence of a legal and regulatory framework in others indicated the GMS countries were not then ready to go ahead with a free electricity market.
2011	The 50 Year Success Story – Evolution of a European Interconnected Grid	UCPTE and UCTE	Progress and technical improvements since the Union for the Coordination of Transmission of Electricity formed in 1951 prior to the evolution of ENTSO-e in July 2009.

ADB = Asian Development Bank, ASEAN = Association of Southeast Asian Nations, ENTSO-e = European Network of Transmission System Operators for Electricity, GMS = Greater Mekong Subregion, HVDC = high voltage direct current, IEA = International Energy Agency, IRENA = International Renewable Energy Agency, JICA = Japan International Cooperation Agency, kV = kilovolt, LFC = load frequency control, MHI = Manitoba Hydro International, MOU = memorandum of understanding, NGC = National Grid Code, NREL = National Renewable Energy Laboratory, RPTCC = Regional Power Trade Coordination Committee, TA = technical assistance, UCPTE = Union for the Co-ordination of Production and Transmission of Electricity, VRE = variable renewable energy.

Source: Authors.